Flowering Plants

Edited by Jo Kitz and Linda Hardie-Scott

Photographs edited by Dede Gilman

Collaboration on historical matter by David Hollombe

Flowering Plants

THE SANTA MONICA MOUNTAINS, COASTAL & CHAPARRAL REGIONS OF SOUTHERN CALIFORNIA

Text by Nancy Dale

Photographs by members of the
California Native Plant Society

Illustrations by Marianne D. Wallace

CAPRA PRESS
1986

FLOWERING PLANTS was designed by Francine Rudesill
with typography by Jim Cook Book Design & Typography.
Cover photo by Tim Thomas.
The four-color plates were prepared and printed by
Dai Nippon Printing Company, Ltd., Tokyo, Japan.

LIBRARY OF CONGRESS CATALOGING-IN-PUBLICATION DATA
Dale, Nancy.
FLOWERING PLANTS.
Bibliography: p. Includes index.
1. Botany—California—Santa Monica Mountains Region.
2. Wild flowers—California—Santa Monica Mountains Region—Identification.
3. Plants—Identification.
4. Santa Monica Mountains Region (Calif.) I. Title.
QK149.D23 1985 582.13'09794'93 85-17461
ISBN 0-88496-239-3

Published by CAPRA PRESS
Post Office Box 2068, Santa Barbara, Ca. 93120
in cooperation with
THE CALIFORNIA NATIVE PLANT SOCIETY

This book was made possible through a grant from the Santa Monica Mountains Conservancy, an agency of the State of California, which was created to acquire lands and provide programs within the Santa Monica Mountains.

All proceeds from this book will be used for educational and interpretative materials within the Santa Monica Mountains National Recreation Area.

ACKNOWLEDGEMENTS

I was helped by many people while writing this book. I cannot begin to list them all, but I am particularly grateful to the following: Geoffrey Burleigh and James Kenney who are wildflower photographers of note and from the very beginning their willingness to donate from their comprehensive collections of flower photos made the concept of a picture book seem possible; David Hollombe who corrected botanical errors and contributed invaluable information about the meanings of the scientific names, a subject he has researched diligently for years; Marianne D. Wallace whose botanical illustrations have appeared in *The Los Angeles Times* and publications of the Los Angeles State and County Arboretum, the Huntington Library, the Los Angeles County Art Museum, and others who graciously and cooperatively donated her services; The California Native Plant Society which gave me early encouragement as well as pledging financial support; Stephen W. Austin of El Monte who was helpful with information about butterflies; Bert Wilson, Las Pilitas Native Plant Nursery, Santa Margarita who shared his knowledge of the successful use of natives in Southern California gardens; and my cousins, Mr. and Mrs. C.A.H. Thomson of Kensington, Maryland who also encouraged and contributed.

To no others, however, am I more grateful than to Jo Kitz, immediate past president, and Linda Hardie-Scott, current president, of the Santa Monica Mountains Chapter of the California Native Plant Society. Without their tireless and unceasing efforts over a period of many months to achieve the necessary financing for publication, this book would not exist.

—NANCY DALE

PHOTOGRAPHERS

Nancy Dale	*Santa Paula*
Geoffrey Burleigh	*San Fernando*
Arline Denny	*Northridge*
Dede Gilman	*Newhall*
Linda Hardie-Scott	*West Los Angeles*
Stanley J. Higgins	*Hawthorne*
Claire Kaye	*Malibu*
James P. Kenney	*Pacific Palisades*
Steven R. Kutcher	*Pasadena*
Sarah Thomas Schwaegler	*Santa Monica*
Barry Silver	*West Los Angeles*
Margaret Stassforth	*Los Angeles*
Suzanne Swedo	*Los Angeles*
Tim Thomas	*Calabasas*
Janet Vail	*Agoura*
Frances Vogel	*Pacific Palisades*

THE CALIFORNIA NATIVE PLANT SOCIETY is an organization of laymen and professionals united by an interest in the plants of California. It is open to all. Its principal aims are to preserve the native flora and to add to the knowledge of members and the public at large. It seeks to accomplish the former goal in a number of ways: by undertaking a census of rare, endangered, and extinct plants throughout the state; by acting to save endangered areas through publicity, persuasion, and on occasion, legal action; by providing expert testimony to governmental bodies; and by supporting financially and otherwise the establishment of native plant preserves. Its educational work includes: publication of a quarterly journal, *Fremontia,* and a periodic *Bulletin;* assistance to teachers and school projects; meetings, field trips, and other activities of local chapters throughout the state. Non-members are welcome to attend meetings and field trips.

The work of the Society is done by volunteers. Money is provided by the dues of members and by funds raised by chapter activities. Additional donations, bequests, and memorial gifts from friends of the Society can assist greatly in carrying forward the work.

CALIFORNIA NATIVE PLANT SOCIETY
909 12th Street, Suite 116
Sacramento, CA 94814

To be able to call the plants by name
makes them a hundredfold more sweet and intimate.
Naming things is one of the oldest and
simplest of human pastimes.

HENRY VAN DYKE in *Little Rivers*

TABLE OF CONTENTS

PREFACE

This book is a popular introduction to the abundant flowering plants of the Santa Monica Mountains. The pictures and descriptions are of native and naturalized species—plants growing wild and untended. More than eight hundred sixty species of flowering plants have been discovered here which include trees, grasses, sedges and rushes, introduced weeds, garden escapes, and quite a few that are beyond the scope of this presentation.

In this book two hundred forty-nine of the most commonly encountered species are written about in detail accompanied by photographs or drawings. One hundred thirty-three more species are also described in the text as well as the fifty-six families to which they all belong. In a separate chapter, sixteen trees are described with leaf illustrations.

The pictures and descriptions should help you recognize the flowering plants around you as you enjoy hikes, strolls, picnics, bike rides, bird walks and other outings in this wilderness which still exists by some miracle in the backyard of Los Angeles.

Botanical terms are avoided whenever possible. Those that have been used, such as "sepal," "calyx," "bract," are defined in a short glossary.

For each flower there will be:

1. A color picture or a botanical drawing.

2. The common English name. The common name of a flower is usually acquired by usage often of long association, but sometimes it can be subject to confusion since there may be more than one common name for the same plant, even in the same area. In many cases a flower will be called something else in another part of the country. The flower lover who grew up in Virginia will *insist* you are not showing him a "Skullcap" because the one he knows so well by that name is not like the one that grows in California. "Bluebells" may mean any of a half dozen different kinds of plants. "Willowherb" is not a willow and "Rock-rose" is not a rose.

3. The scientific name of the genus and species. It is unfortunate that many readers are alarmed by the foreign names, usually Latin, that botanists have given plants. Some feel that it is pretentious to use them. The disagreements and inaccuracies that occur with common names are replaced by the precision of scientific names. The Latin insures that those speaking Japanese, German or Hungarian will know what plant is meant.

The scientific name of a plant always consists of two names: the first is the genus and the second is the species. The genus refers to a group of plants that have similar characteristics such as the lily. The genus for the lily is *Lilium.* (The genus name is always capitalized.) Following the genus name is the species name. Species are composed of similar individuals able to produce others of like kind. There are many lilies in the genus *Lilium* but the one often called Lemon Lily is *Lilium parryi.* (Species names are not capitalized. The two-part scientific name is printed in italics, or if this is not possible, the name should be underlined.) One or more names or initials follow the scientific name. These designate the authority who gave the plant its name.

4. The common and scientific name of the family. Genera with similar characteristics are grouped under a family name that usually ends in ACEAE and is generally taken from the name of one of its genera. *Lilium parryi* is a species of the genus *Lilium* belonging to the Lily Family, LILIACEAE.

A brief description of some of the more prominent characteristics is given for each of the families in the text.

5. Blooming and location. Flowering time and the usual habitat are given—where and when you can expect to find these flowers in the Santa Monica Mountains. Some plants thrive on sunny roadbanks and others need to keep their "feet" wet. Knowing the kind of situation a flower must have can often aid in its identification. If you are standing on a dry rocky ledge you are not likely to see a flower found only on a sandy beach. Try to understand the plant communities listed in the introduction so that if you are told that a flower is "abundant in Coastal Sage throughout" you will know where you and the flower both are.

"Chocolate Lilies have been right *here* on April 10 for six years, so why aren't they here now?" cries the native plant buff more often than you would suppose. But, plants as living organisms are not as predictable as we would like them to be, so variability of precipitation, temperature, fires and other imponderables make time and location only approximate. Furthermore, bulldozers, hungry rabbits and encroaching weeds prevent flowers from appearing where they have been reported.

This book is intended as a guide. If you are determined to find a particular plant try a similar habitat or a different time next year.

6. Interesting information is given about each particular plant—how it was used by Indians or early settlers, if it is edible, if it is useful in the garden, if it has an unusual method of pollination or has a special smell or a well-known garden relative.

However, there are two distinct warnings that must be emphasized. It is against the law to cut, pick or dig any native plant in the entire state unless you are on your own property or on private land with the permission of the owner. The botanists who collect for scientific purposes have established their right to do so and have official permits.

The second caution is for your own safety. Many of the attractive plants you will see in these mountains are quite toxic in all or in part and identification is never certain, particularly for beginners. The information about Spanish or Indian use is for interest only and is not intended in any way to encourage you to duplicate their diet or their pharmacopoeia. A few recipes are included. Be sure you know your plants and have permission to gather.

7. What the scientific name means. How plants get their names is a fascinating story in itself. Perhaps you'll find the unfamiliar Latin more acceptable if you know that the words often say something important about the appearance of the plant—the leaves are in little bundles *(fasciculatum);* maybe the stem is smooth *(laevicaulis)* or covered with hair *(hirsutus);* the petals might be velvety *(velutinus),* striped *(vittatus)* or yellow *(luteus). Medicago* goes back through the centuries when "alfalfa" first came to Greece from Medea and was called "medice." *Ranunculus* in Latin is "little frog" and most of the many species of this genus like moist habitats, including our buttercups. *Helianthus* comes from two Greek words meaning "sun" and "flower" and is the genus of the common sunflower.

Some names honor the scientist who first collected the plant. A first collector can choose to name it in honor of a good friend or sponsor. Examples are *Fremontodendron* for John C. Fremont, a western explorer, or *douglasii* for the early Scottish botanist, David Douglas. Sometimes the name is an indication of where the plant was first found—*californica* and *mexicana* are obvious.

THE PARTS OF A FLOWER

The flower is the reproductive part of the plant, the part that produces the seeds that produce new plants. Sometimes a flower is small and inconspicuous, but often it is large and colorful, adapted to attract the insects that are necessary for fertilization.

The wide differences among flowers are most apparent in the outer parts, the petals and sepals. The sepals cover all of the other parts in the bud.

The similarities are most noticeable in the inner parts, the pistil and stamens. In a fairly typical flower the pistil, consisting of ovary, style and stigma, is situated in the center of the receptacle that terminates the stalk. The ovary contains minute ovules, the future seeds. Several stamens stand around the pistil. The anthers at the top of the stamen contain chambers in which the pollen, a mass of microscopic living parts, is formed. The anthers open, freeing the pollen to be transported to the stigma of a pistil. If conditions are compatible, each pollen grain may send forth a slender thread which penetrates the stigma, proceeds down inside the style and enters first the ovary and then an ovule. This union results in a new cell which develops into an embryo, the small plant within the seed. As the seed grows the ovary changes in many ways, particularly in size. In many flowers, the surrounding parts that join with the ovary also undergo great change. Botanists call this body enclosing the seed the fruit.

Thus, it can be seen that the chief business of the flower is reproduction.

Petals and sepals may be fused together in forms that look like bells, tubes, funnels or urns. Counting parts can help you identify the flower you are examining. Don't expect all parts to be present in all flowers. There are endless variations. Wind-pollinated flowers may lack petals and have their stamens on one plant and the pistils on a next-door neighbor.

All of the petals, sepals and stamens of each regular flower radiate from the center. Those that vary from this arrangement are called irregular. The blossoms of peas, snapdragons and mints are irregular.

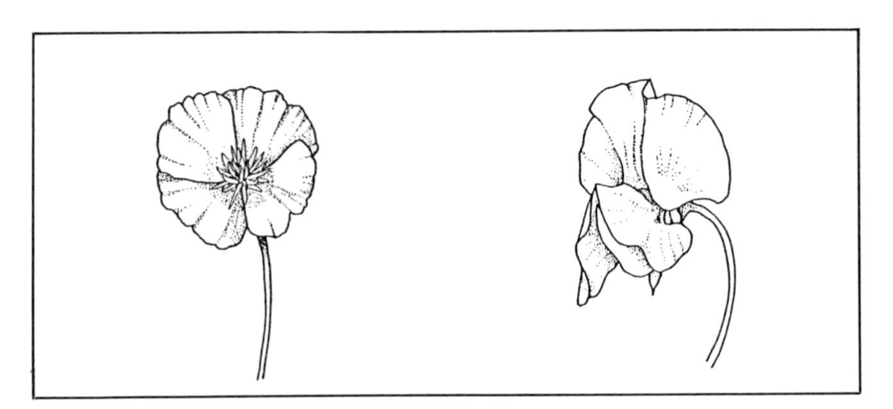

Example of a Regular Flower

Example of an Irregular Flower

One of the largest plant families in the world is the ASTERACEAE, or COMPOSITAE as it is also called. A "flower" of this family is actually made up of many individual flowers held together in a single head and looking like one bloom. Some, like the thistle, have only a tight little bunch of tubular flowers called disk florets in the center of the head. Others, like the dandelion, have only flowers with strap-like petals which are called ray florets. But many have both, as in this diagram.

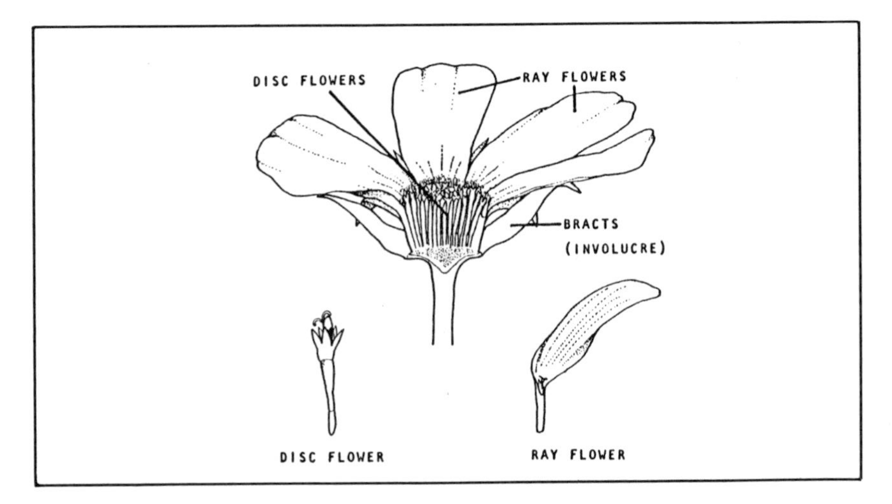

Composite Flower

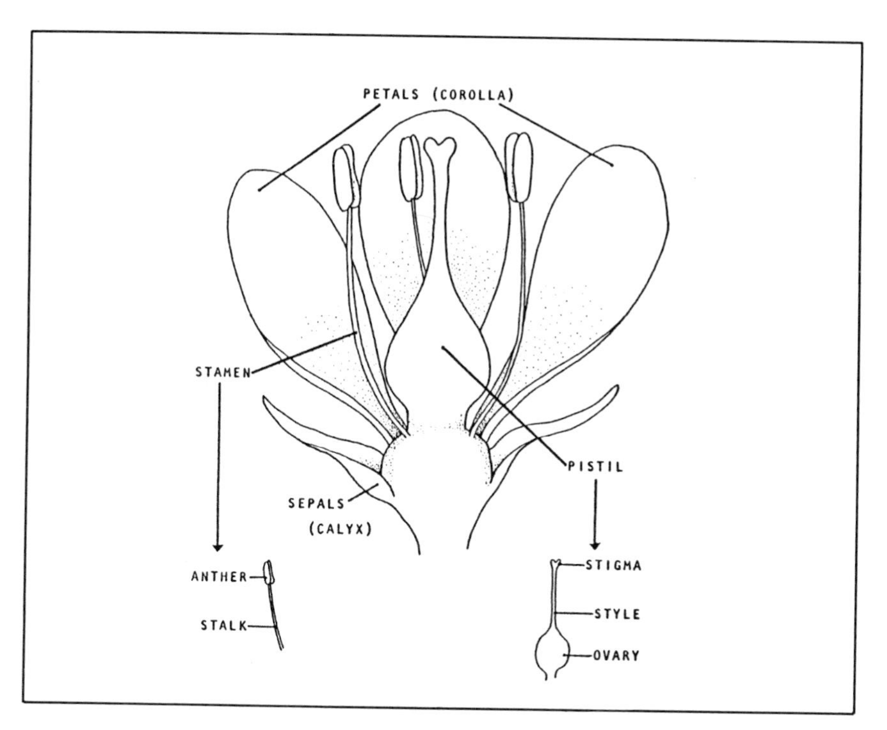

Flower

Another aid to correctly identifying a plant is an examination of leaf shapes and arrangements. Learn to differentiate between these three arrangements.

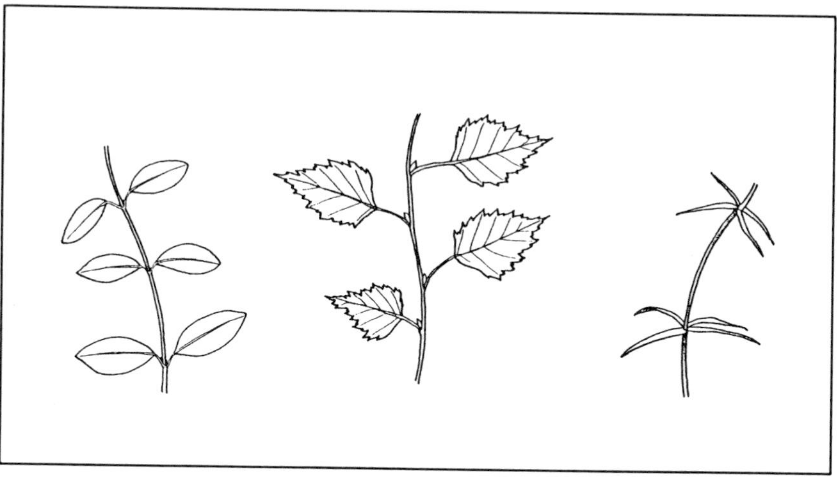

Opposite Leaves *Alternate Leaves* *Whorled*

Some leaves are composed of "leaflets" which may be hard to distinguish from a true leaf. These are called compound leaves and it will help to learn the two terms illustrated below.

Pinnately Compound *Palmately Compound*

Since some of the plant parts you will be examining are very small you will find using a hand lens, eight or ten power, very helpful.

If you get very serious and begin to use a more technical book than this one, you will want a pair of needle picks and a razor blade to "operate" on the flower you are examining—to open the tiny parts for your inspection.

Most floras organize plants into "monocots," which have one seed leaf, and "dicots," which have two seed leaves. This book is no exception. The division is based on whether the emerging plant after germination has at the very first one little leaf or two. These first emergents are called seed leaves and monocotyledon means one leaf and dicotyledon means two leaves.

When fully grown with many leaves it is still very easy to determine a monocot from a dicot. If the veins on the leaf are parallel and never touch you have a monocot. Orchids, corn and grasses are monocots. Their flower parts are mostly in three's or sixes. AGAVACEAE, AMARYLLIDACEAE, IRIDACEAE, LILIACEAE, ORCHIDACEAE are the families of monocots in this book and they come first.

All of the others are dicots which have networks of veins in their leaves, and their flowering parts are usually in fours or fives.

AN INTRODUCTION TO THE SANTA MONICA MOUNTAINS

The Santa Monica Mountains in the Los Angeles area are remarkable for many reasons, not the least of which is that they are the only mountain range in the country to bisect a great city. Ten miles across at its widest point and fifty miles long from Griffith Park to Point Mugu, this range lies within an hour's drive of more than ten million people. In its many canyons and on its ridges it preserves thousands of acres of wilderness in the midst of an urban area. To protect this magnificent natural area from further development, Congress, in 1978, created the Santa Monica Mountains National Recreation Area. This provided for a 70,000-acre park which was to be managed by the National Park Service to be added to the already existing 33,000 acres of state park land. Some of this acreage has been acquired.

Within the state park system in the Santa Monica Mountains are three main parks, Topanga, Malibu Creek, and Point Mugu, as well as substantial state beach holdings. One of the largest of the beach parks, Leo Carrillo, contains extensive areas of Chaparral and Riparian Woodlands. Griffith Park, created in 1896 when Colonel Griffith J. Griffith gave Los Angeles 3000 acres of his ostrich ranch, is included. City parks and two county parks, Charmlee and Tapia, bring public holdings at present to less than one-fifth of the 200,000 acres covered by the range.

Geologists tell us the Santa Monica Mountains are quite young, but their structure is complex. They have been uplifted and then eroded several times over a period extending back nearly two hundred million years. They are generally uplifted, faulted, folded and east-west trending.

Linda Hardie-Scott

Big Sycamore Canyon *Point Mugu State Park*

Some of the oldest granites can be observed where they have been exposed by erosion in Griffith Park. The dark Santa Monica slates can be seen in Topanga State Park from Temescal to Franklin Canyon, and in the roadcuts along the San Diego Freeway. These slates were originally formed when mud or shale was deposited beneath a shallow sea some one hundred forty million years ago.

About one hundred twenty million years ago molten rock intruded baking the shales into slate and cooling into granite. After much uplifting and erosion, seas again covered the land depositing the marine rocks that form the great cliffs in lower Topanga Canyon.

About sixty million years ago there was yet another cycle of erosion and returning of the sea resulting in what geologists call the Martinez Formation. This can be seen in Topanga Canyon also. The Sespe Formation, easily recognized by its predominant red-purple color, followed the erosion of the Martinez Formation and can be seen in the sea cliffs along the Pacific Coast Highway and in Malibu Canyon.

About fourteen million years ago the sea again covered the central part of the Santa Monicas depositing thick-bedded sandstones, coarse-grained and buff-colored. At this time there was also extensive volcanic activity which established a north-south trend that has since been rotated clockwise 90 degrees and compressed into the present east-west trend of the Transverse Ranges.

Brown or reddish volcanic rock can be seen in many places in the western part of the range. Modelo shale which forms a mantle over the older rock layers on the north flank is the youngest formation in the Santa Monicas. It is light-colored, bedded and commonly contains fish and seaweed fossils.

Beginning about three million to ten thousand years ago the Santa Monicas began to be uplifted again and geologists believe the Malibu coast is still rising.

Topanga and Malibu Creeks have cut gorges hundreds of feet deep on their way to the Pacific. Malibu Creek, originating in the Simi Hills, is thought to have flowed in its present course before the mountains existed but was able to cut down fast enough to maintain its position through the ages. Topanga Creek is probably an old stream also. It begins its journey at the top of Topanga near the north flank of the range.

The soils of the area range from thick well-drained loams of gently sloping valley bottoms supporting groves of oaks to thin rocky soils of the steep hillsides where chaparral plants thrive.

The range is ten miles wide at its broadest part and reaches an elevation of 3111 feet at Sandstone Peak near the western end. The altitude of the major portion rises between 1000 to 2000 feet.

The mild rainy winters and hot dry summers constitute a typically Mediterranean climate. The slopes facing the ocean benefit from a cooling sea breeze and are never as warm as the interior which can be uncomfortably hot in the summer. There is an occasional frost (rarely below 25 degrees) away from the coast and once every twenty years or so a light snow falls, disappearing almost before the photographer can get film in the camera. There is a dry season from May to October and a distinct rainy season from November to April, but the precipitation is limited to relatively short periods. There are only a few days in the year when outdoor activities cannot be enjoyed. The mountains are at their best in late winter and early spring when the temperatures are cool and the rains bring out lush stands of wildflowers.

There is a great variety of animal life from vociferous frogs and scurrying lizards to rarely seen mountain lions. Hikers often see ground squirrels, rabbits, coyotes and mule deer, but the opossums (introduced), raccoons, skunks, foxes and bobcats are not so frequently spotted. The rattlesnake, much shyer than commonly supposed, is never seen by most visitors and rarely is anyone bitten.

Since this area is in the Pacific flyway many transient birds may be sighted as well as soaring vultures, hawks, golden eagles, peregrine falcons and even an occasional giant condor. The large waterbirds—pelicans, great blue herons and ospreys—have been seen in the southern areas.

The variety that marks Southern California's natural setting is nowhere more evident than here in the Santa Monicas. There are shady chasms, waterfalls plunging among fern-draped boulders and chaparral-covered ridge-tops affording panoramas of higher ranges to the north or the sunlit sea to the south.

At least nine different plant communities flourish in the Santa Monicas beginning with the rich kelp beds offshore. Miles of beaches, more nearly pristine than might be expected considering the extent to which they are in use, exhibit Coastal Strand species such as Sand Verbena (*Abronia* spp.) and Beach Primrose *(Camissonia cheiranthifolia).* The Salt Marshes at Malibu Lagoon (ten acres at the mouth of Malibu Creek) and many more acres at Point Mugu support typical plants such as Pickleweed *(Salicornia virginica),* Seepweed *(Suaeda californica)* and Sea Lavender (*Limonium* spp.)

Linda Hardie-Scott

Coastal Sage Scrub *Point Mugu State Park*

Coastal Sage Scrub is a plant community that occupies the western slopes above the beaches. Coastal Sagebrush *(Artemisia californica)* and three species of Eriogonum *(E. cinerea, E. elongatum* and *E. fasciculatum)* characterize this association as it climbs the slopes away from the sea along with other more showy herbs such as various species of *Lotus, Lupinus* and *Mimulus.*

Linda Hardie-Scott

Chaparral *Topanga State Park*

An almost unique association of plants has also evolved on these mountains. This is known as Chaparral and is widespread on the inland slopes and ridges. Here Chamise *(Adenostoma fasciculatum),* California Lilacs (*Ceanothus* spp.), Scrub Oak *(Quercus dumosa),* and Toyon *(Heteromeles arbutifolia)* are adapted to winter rains, hot dry summers and frequent fires.

Linda Hardie-Scott

Riparian Woodland *Santa Ynez Canyon*

On the shallow margins of ponds and little lakes small Freshwater Marshes are found with sedges (*Carex* spp.), tules (*Scirpus* spp.) and cattails (*Typha* spp.) Along the banks of the permanent streams appear Riparian Woodlands—some of the loveliest areas in the mountains—with tall Western Sycamores *(Platanus racemosa),* Fremont Cottonwoods *(Populus fremontii)* and California Bays *(Umbellularia californica).*

Sarah Thomas Schwaegler

Oak Woodland *Malibu Creek State Park*

Well-developed Oak Woodlands are found in Topanga Canyon, Malibu Canyon and the Hidden Valley area. Coast Live Oak *(Quercus agrifolia)* dominates that part of the community that has a rich understory of ferns and shrubs. In some spots the Bracken Fern grows to 11 feet tall and is believed to be the largest in the United States. Valley Oak *(Quercus lobata),* grows here at its southernmost limit in the open grassy areas called Oak Savannas.

Linda Hardie-Scott

Grassland *Point Mugu State Park*

There are Grasslands, but they are not extensive. Where they do exist they are composed largely of introduced annual grasses. However, some native perennial bunch grasses, notably *Stipa pulchra,* have survived and are particularly common in the La Jolla Grassland in Point Mugu State Park.

Five species listed in the California Native Plant Society's *Inventory of Rare and Endangered Plants of California.* (3rd ed., 1984) are located within these mountains.

Dudleya cymosa subsp. *marcescens* is located on Malibu Beach, Point Dume and the Triunfo Pass. It is extremely rare and found only in California. The common name for this little succulent is the Santa Monica Mountains Live-forever. It is difficult to locate even when in bloom in May and June.

In 1978 the Santa Susana Tarweed *(Hemizonia minthornii),* long thought to exist only in the area of Santa Susana Pass, was discovered in Corral Canyon. This attractive yellow composite thrives in sandstone outcroppings and seems to grow out of solid rock. It can be quickly identified by its yellow anthers. *H. minthornii* is extremely rare and found only in the Simi Hills (type locality) and the Santa Monica Mountains. It blooms from July to October.

Lyon's Pentachaeta *(Pentachaeta lyonii),* a beautiful little low-growing composite, blooms in March and April in the grassy areas in the western end of the mountains. It is also very rare and a California endemic.

The Conejo Buckwheat *(Eriogonum crocatum)* is the fourth rare and endangered species. It is not quite as rare as the preceding ones but is restricted to the rocky slopes of the Conejo Grade and at one station near Westlake. This perennial subshrub covers itself with sulphur yellow balls of flowers beginning in April and lasting until July. It is most handsome.

Salt Marsh Bird's Beak *(Cordylanthus maritimus* subsp. *maritimus)* can be found on United States Navy land at Point Mugu blooming from May to October. Most of the flowers of the genus *Cordylanthus* are as neutral in color as the grasses surrounding them, but the flowers of this one are a glorious pink and are spectacular to see. A field of rosy bloom! It is nice to know that the Navy is aware of its presence and its significance and is concerned for its future.

Recently the State of California and the Federal Government have enacted rare plant laws and have placed significant plants under protection. The State listed over thirty, which included the Conejo Buckwheat, the Dudleya, the Tarweed and the Pentachaeta. The Federal Government so far has listed thirteen for the entire country and one of those was the Bird's Beak. The Pentachaeta is being strongly considered for federal protection as well.

The Santa Monica Mountains are unique for many reasons, not the least of which is their exciting and unusual flora. It is certainly hoped that this great national resource will remain undeveloped enough to insure their continued colorful blossoming.

Flowering Plants

MONOCOTS

Agave Family. AGAVACEAE

Agaves are woody plants with long, stiff, sword-like leaves usually crowded near the base of the stem. Flowers have 6 petals, 6 stamens and a 3-lobed stigma. The fruit is a capsule or a berry.

Prior to the recognition of this family, its members were placed in the LILIACEAE or in the AMARYLLIDACEAE.

Yucca whipplei is the only species of this family in the Santa Monicas.

Our Lord's Candle
Yucca whipplei Torr. ssp. *intermedia* Haines.
Agave Family. AGAVACEAE

Geoffrey Burleigh

The flower stalk is 4 to 8 feet tall. There is a basal rosette of spine-tipped, sword-like leaves 1 to 3 feet long. The white flowers, sometimes purple-tinged, appear in a terminal compound cluster up to 4 feet long. They are very fragrant and even have a pleasant taste when nibbled. The fruit is a capsule.

Beginning in April and continuing through July, Our Lord's Candle blooms abundantly in Chaparral and Oak Woodland in elevations up to 2500 feet.

There are 40 species in the genus *Yucca*, but this is our only one. The shoot bearing the flower dies after flowering, but produces many offsets around the old roots. Indians made flour of the seeds and used fibers from the leaves to weave cordage, nets and baskets. They roasted very young flower stalks to produce a food that tasted like baked apple. The roots have a high component of saponin; when soaked and pounded they produce copious suds.

The relationship between the yucca moth and the yucca plant is a classic example of symbiosis. The California Yucca Moth *(Tegeticula muculata)* is the one that interacts with *Yucca whipplei.* The larvae feed only in yucca fruits that only develop from moth-pollinated flowers. The female flies to a flower at night, rolls a ball of pollen bigger than her head and then carries it in her mouth to another flower where she makes a hole in the pistil, deposits her eggs and then places the ball of pollen on top of the pistil. When the eggs hatch, the caterpillars eat a few seeds, make a hole in the pod, come out, spin a thread to lower themselves to the ground where they bury themselves, then emerge in about a year as moths.

Yuca is the native name for *Manihot*, a genus in the Spurge Family (EUPHORBIACEAE) which is commonly called Sweet Cassava or Tapioca. A.W. Whipple (1816-1863), a topographical engineer (surveyor), commanded the Pacific Railroad Survey to Los Angeles in 1853.

Amaryllis Family. AMARYLLIDACEAE

These are perennial, herbaceous plants closely related to the Lily Family, the chief difference being that the flowers of the AMARYLLIDACEAE appear in umbels, subtended by clasping bracts on a leafless stem. Leaves and flowering stalks grow from rhizomes, corms or bulbs.

The leaves are basal. There is one style and a 3-lobed stigma. The fruit is a capsule or a berry.

In the older floras, the genera in this family will be listed as belonging to the LILIACEAE. We have one species each of *Bloomeria, Brodiaea, Dichelostemma,* and three of *Allium.* The flowers of all of them are so lovely and so often encountered that most of them are pictured here.

Peninsular Onion

Allium peninsulare Lemmon.
Amaryllis Family. AMARYLLIDACEAE

Geoffrey Burleigh

From an oval, brown bulb about ½ inch long, a smooth, leafless stalk arises, topped by a cluster of 6 to 25 shining red-purple flowers on short stemlets subtended by short papery bracts. Each flower has 6 petals about ½ inch long. The 2 to 4 narrow, strap-like leaves are only ½ inch wide. The fruit is a capsule. The members of the genus *Allium* always have the taste and smell of onions.

A. peninsulare blooms from April to June on rocky slopes in Grassland and open Oak Woodland. It has been found in Griffith Park, near Lake Sherwood and on the Conejo Grade.

If you find a flower resembling this one but pale pink to white, you probably have the Red-skinned Onion *(A. haematochiton)* which is another species found in the Santa Monicas. They are both as edible as the cultivated onions, but far too attractive for such a fate.

Allium is Latin for "garlic." *Peninsulare* refers to the first collection of the species in Baja California, a peninsula. *Haematochiton* is from Greek for "blood and coat," referring to the color of the skin of the bulb.

Golden Stars
Bloomeria crocea (Torr.) Cov.
Amaryllis Family. AMARYLLIDACEAE

Linda Hardie-Scott

The plant arises from a corm. The bare stem is 6 to 14 inches long. There are 2 grass-like leaves at the base. The 30 to 50 star-like, yellow-orange flowers appear in a cluster on little stalks all arising at the top of the stem. The petals have a brown stripe down the middle.

The difference between a *Brodiaea* and a *Bloomeria* can be determined by looking at the stamens of the latter. They rise from a tiny cup at their base. This little cup is not present in *Brodiaea.* The fruit is a capsule.

B. crocea can be found from April through June in Oak Woodland and disturbed Chaparral away from the coast.

Golden Stars seem to make fields twinkle when spotted in the browning grasses of late spring. Although it is not well known in home gardens it is often seen in botanic gardens where it volunteers abundantly. It is propagated from seed, requiring three to four years to become a mature bulb. Golden Stars prefer full sun and a semi-dry condition.

H. G. Bloomer (1821-1874) was an early San Francisco botanist and one of the founders of the California Academy of Sciences. *Crocea* means "saffron-colored."

Harvest Brodiaea
Brodiaea jolonensis Eastw.
Amaryllis Family. AMARYLLIDACEAE

Geoffrey Burleigh

The slender, leafless stem is 7 to 20 inches tall topped by 3 to 11 trumpet-shaped, blue-violet flowers ¼ to 1¾ inches long on 1 to 3½ inch stemlets. The narrow leaves appear at the base. The 3 fertile stamens alternate with 3 sterile ones. The fruit is a capsule.

Harvest Brodiaea blooms in April and May in dry clay soil on the northern slopes of the central part of the mountains. Of the 40 known species, 29 are native to California, but this is the only one in our mountains and is not abundant.

The solid bulb is edible and was a favorite tidbit of the Indians who roasted it slowly in hot ashes. Almost all of the Brodiaea bear many seeds in papery capsules. These seeds germinate readily and produce mature corms in about three years.

The generic name honors James Brodie (1744-1824), a Scottish botanist who specialized in algae, ferns and mosses. The original specimen was collected in Jolon, California.

Blue Dicks, Wild Hyacinth
Dichelostemma pulchellum (Salisb.) Heller.
Amaryllis Family. AMARYLLIDACEAE

Dede Gilman

Blue Dicks have a bare stem 1 to 2 feet tall and grass-like leaves that can be up to 16 inches long. There are 4 to 10 purple-blue flowers in a head-like cluster. There are 6 unequal anthers, 3 of which have a wing extending up into two appendages which hide them. This feature removed this plant from the genus *Brodiaea* where you may find it listed in older floras. The fruit is a capsule.

Blue Dicks, or Wild Hyacinth, are very common in Coastal Sage and Grassland throughout, blooming February to May.

The little corms are tasty and were enjoyed by the Indians and the children of early settlers who called them grass nuts. The blossoms keep a long time when gathered.

The generic name derives from two Greek words meaning "a garland which is twice-parted to the middle." This refers to the appendages on the stamens. *Pulchellum* is Latin for "beautiful."

Iris Family. IRIDACEAE

Plants in the Iris Family are perennial herbs with parallel-veined, narrow, sheathing leaves folded lengthwise to enclose the next one within. This feature was fancifully called "equitant" by Linnaeus—riding horseback. The flowers are showy. The ovary is inferior. There are only 3 stamens. The fruit is a triangular capsule that opens when mature to disperse the seeds.

There are about 1200 species of this family in the world, and you will know many of them as beautiul garden ornaments. We have only two native genera in the state, *Iris* and *Sisyrinchium.* Only the latter is native in our mountains.

Blue-eyed Grass
Sisyrinchium bellum Wats.
Iris Family. IRIDACEAE

Nancy Dale

The deep blue flowers have a yellow center and are 6-parted. They appear at the top of branching flattish stems which are a foot or so high. The leaves are basal and grass-like, although they do exhibit the clasping that is a feature of the whole family. The fruit is a triangular dry capsule.

Blue-eyed Grass is common in Grasslands and Coastal Sage throughout. It blooms from March through June.

The Spanish Californians called *S. bellum* "azulea" and made the roots into a tea to use when feverish. A patient was supposed to be able to survive on this tea for many days with no other sustenance.

When planted in full sun *S. bellum* will grace your garden with increasing clumps of bloom from early to late spring.

The name goes all the way back to Theophrastus (372-287 B.C.), a Greek student and successor of Aristotle whose works on plants were heavily relied on by Renaissance writers. *Sisyrinchium* refers to a plant he described related to the *Iris. Bellum* means "handsome."

Lily Family. LILIACEAE

Plants of the Lily Family are usually perennial herbs with parallel-veined leaves. The flower parts are in threes or sixes with petals and sepals often alike in color. There are six stamens and a superior ovary. The fruit is either a dry capsule that opens when mature or a fleshy berry.

Some of the most beautiful and fragrant garden flowers—hyacinths, lilies, tulips—belong to this family. Our native members are also very lovely—the truly magnificent Humboldt Lily and the six exquisite Mariposas *(Calochortus).* The *Calochortus* are the most widely dispersed of all the native liliaceous plants of the Pacific Coast and comprise some of the most beautiful flowers of the world.

Unlike most of the other members of the Lily Family, the petals and sepals of the *Calochortus* are not alike in color, size or shape. There are three types of *Calochortus:* the globe type of which we only have the Fairy Lantern *(C. albus);* the Star Tulips or Cat's Ears of which we have none; and the Mariposas with fan-shaped petals of which we have the five described here.

Fairy Lantern, Globe Lily

Calochortus albus Dougl. ex Benth.
Lily Family. LILIACEAE

Linda Hardie-Scott

The smooth stems are 1 or 2 feet high. The strap-shaped leaves are ½ to 2 inches wide. There are 3 small greenish sepals and 3 large white petals. The satiny, white flowers are nodding and the petals arch to form a closed globe which is hairy within.

Fairy lanterns are common in deep shade in Oak Woodlands on north slopes in the central part of the range. They bloom in April and May. One flower book says "doubtless God could have made a lovelier flower, but never did."

Although this is the only species of the globe form in our mountains, farther north in California two glowing-yellow cousins, *C. pulchellus* and *C. amabilis,* are found. *C. amoenus,* a two-toned pink globe beauty, grows abundantly in Fresno County and *C. albus,* var. *rubellus,* a rose-red variety of *C. albus,* grows in San Luis Obispo County.

Calochortus means "beautiful grass" in Greek. *Albus* means "white."

Catalina Mariposa Lily
Calochortus catalinae Wats.
Lily Family. LILIACEAE

Geoffrey Burleigh

The flowers are erect, 2 to 3 inches across and are terminal on 1 to 2 foot stems. They are white to pale lilac and are darkened with a red blotch at the base of each petal. The leaves are grass-like.

C. catalinae is common in Grassland and Coastal Sage at low elevations throughout. It blooms in March, April and May.

Mariposa means "butterfly" in Spanish. When you see the lovely flowers hovering above the grass on almost invisible stems and dancing in the slightest breeze, you will know how apt the name is. You will be enchanted with the sight—unless, of course, you are trying to take a picture.

The species was first described from a specimen collected on Catalina Island.

Yellow Mariposa Lily
Calochortus clavatus Wats.
Lily Family. LILIACEAE

Geoffrey Burleigh

Calochortus clavatus is a tall plant—up to 40 inches—with stiff, zig-zagging stems and narrow, grass-like leaves which are usually dried and gone by flowering time. Bright yellow petals—up to 2 inches long—form a broad-based cup showing within a ring of club-shaped hairs. Dark purple anthers radiate star-like in the center. Both the flowers and the dry fruit container are erect. As in all members of this genus, the petals and sepals are in 3's.

It is one of the glories of the Santa Monicas that the Yellow Mariposa Lily blooms so spectacularly and abundantly on dry slopes, particularly after a fire. It can be found in bloom from April to June.

Some authorities consider *C. clavatus* to be the largest-flowered and stoutest-stemmed of all the *Calochortus* in California.

Clavatus, "club-shaped," refers to the top of the hairs on the inner side of the petals.

Pink Mariposa Lily
Calochortus plummerae Greene.
Lily Family. LILIACEAE

James P. Kenney

The inner petals of this Mariposa are densely covered with a fringe of yellow hairs. The stems are 1 to 2 feet high. The pinkish-rose flowers are bell-shaped. The basal leaves which have withered by flowering time are up to 16 inches long and ½ inch wide.

C. plummerae is scattered on dry, rocky slopes and brushy areas below 5000 feet. It blooms from May to July.

All of the species of *Calochortus* have a slightly depressed area near the base of each petal. This area on *C. plummerae* is round and bare and surrounded by many silken hairs.

Edward Greene, a 19th century California botanist, described and named this species after the maiden name of John Gill Lemmon's wife, Sarah Allen Plummer (1836-1923), who was a botanist and an expert on ferns and seaweeds. John Gill Lemmon (1832-1908) was a pioneer botanist in Sierra Valley in Sierra County, California.

Splendid Mariposa Lily
Calochortus splendens Dougl. ex Benth.
Lily Family. LILIACEAE

Geoffrey Burleigh

In *C. plummerae* the petals are hairy and the depressed "spot" is bare, but in *C. splendens* the petals are smooth and the "spot" is thick with fungus-like hairs. The petals are a soft lilac color and may have a purple blotch. The basal, strap-like leaves are only ¼ inch wide.

All of the flowers of the genus *Calochortus* are handsome, but none more so than this lavender beauty which can be found on brushy slopes blooming from May to July—but never abundantly.

The Native Americans dug the bulbs of all the *Calochortus* species with a fire-hardened stick and ate them raw or roasted them in rock-lined earthen ovens. The bulbs are reported to be nutritious with a sweet, nutty flavor. This was acceptable when Mariposas were plentiful and people were few. It is absolutely impossible today.

Splendens means "splendid" and certainly fits this species.

Butterfly Mariposa

Calochortus venustus Dougl. ex Benth.
Lily Family. LILIACEAE

Dede Gilman

This is a spectacularly variable species. Many of them have cream to lavender flowers with blotches on the petals that are equally diverse. The flowers have been found colored dark maroon and even pale yellow, striped, spotted or with no markings at all. It is impossible to rely on color for identification. More reliable characteristics are the indented spots on the petals which are large, square and covered with hairs, combined with sepals which are curled back from the tip.

Butterfly Mariposa is found in sandy soil on rocky slopes in the western part of the range blooming from May to July.

All of the species of *Calochortus* can be cultivated for garden use but germination is slow and uncertain and blossoming unpredictable.

Butterfly as a common name for this flower undoubtedly arose from the colorful markings often seen on the petals.

Venustus means "charming." Every flower in this genus is that and none more so than this one.

Soap Plant, Amole
Chlorogalum pomeridianum (DC.) Kunth.
Lily Family. LILIACEAE

Geoffrey Burleigh

The small, white, star-like flowers with 6 recurving parts appear along the top of a leafless 5-foot stem. The flowers open in the late afternoon and are not seen nearly as often as the crinkly wavy-margined foliage of inch-wide basal leaves which stretch in distinctive ripples as far as 18 inches flat along the ground. The fruit is a capsule.

Soap Plant is common throughout the Coastal Sage Scrub and Chaparral. It blooms in May and June.

The bottle-shaped bulb of this plant hides deep in a fibrous jacket. Early settlers used this fiber to stuff mattresses and the Indians used it for brushes. The Indians also crushed the bulb and rubbed it into a lather. They threw it into streams and ponds to stun fish which then floated to the surface and were easily caught. (This practice is illegal today.) They cooked the bulbs and young shoots in an earthen oven and glued their arrows with the thick juice that oozed out.

The Spanish Californians called it *"Escobeta,"* little broom. They also called it *Amole*, a name that comes from the Aztecs.

The genus takes its name from two Greek words meaning "green" and "juice." *Pomeridianum* means "afternoon," when, if you are lucky, you may see the flowers open.

Chocolate Lily
Fritillaria biflora Lindl.
Lily Family. LILIACEAE

Dede Gilman

The nodding flowers are bell-shaped, dark satiny-brown, lined with green and purple inside. The leaves are lance-shaped, somewhat whorled near the base, or scattered on a stem from 6 to 18 inches high. The fruit is a capsule.

Chocolate Lily is local on clay slopes, often under shrubs and blooms from February to April.

This has been called the Cleopatra of the Fritillaries—the darkest and the loveliest. It is our only species, although there are 16 California natives in this genus. They could grace a lightly shaded rock garden or raised border if you were willing to give them special care and wait 4 to 5 years for a mature bulb.

Fritillaria comes from Latin and means "dicebox," referring to the shape of either the capsule or the flower. *F. biflora* often has two blooms to a stem, so *biflora* became the species name. However, plants have been found with over 20 blooms on a stem.

Humboldt Lily
Lilium humboldtii Roezl & Leichtl. var. *ocellatum* (Kell.) Elwes.
Lily Family. LILIACEAE

Linda Hardie-Scott

This tall, stout plant grows from 3 to 12 feet tall. The wavy-margined leaves arranged in 4 to 8 whorls with 10 to 20 in each whorl are 3½ to 5 inches long. The showy, orange-yellow flowers with maroon spots, often incorrectly called the Tiger Lily which it resembles, nod at the end of its tall stem. There are 6 segments, each one over 3 inches long and rolled back and under in a most distinctive way. The underground bulb is very large. The fruit is a capsule.

The Humboldt Lily is common along shaded streams away from the coast. It blooms from May to June.

A specimen was once found that had 50 buds and blossoms, 20 of which were open all at the same time. You may never see such a candidate for the record book but any one of these handsome plants you do find will astound you with its size and beauty.

Lilium is from the Greek *lirion*, "a lily." On his 100th birthday, in 1869, this spectacular flower was named for Baron Alexander von Humboldt (1769-1859), a German geographer who explored Central and South America from 1799-1804.

Star Lily, Fremont Zigadene

Zigadenus fremontii Torr. var. *fremontii.*
Lily Family. LILIACEAE

Dede Gilman

The flowers are star-like, ½ inch in diameter. The 6 spreading segments each have a greenish spot at the base. There are few to many blossoms at the top of the stem which may be up to 3 feet high. The leaves are grass-like and mostly basal with a few short ones scattered along the flower stem. The fruit is a capsule.

The Star Lily is local throughout on dry brushy slopes in Coastal Sage Scrub and Chaparral away from the coast. It blooms from March to May.

Death Camas *(Z. venenosus)* with smaller flowers and narrowly folded leaves is a close relative, but this poisonous species does not occur in the Santa Monicas. It is the bulbs of Death Camas that are deadly when ingested. It is said that pigs are not affected and even like it, so it is sometimes called Hog's Potato.

Zigadenus is from Greek and means "yoked glands." Lt. Colonel John Charles Fremont (1813-1890) was an officer in the American army and presidential candidate who played a prominent part in the early history of California. He collected plants on his military expeditions and sent them to John Torrey, a New York botanist. One plant genus *(Fremontodendron)* and several species' names *(fremontii)* honor his botanical activities.

Orchid Family. ORCHIDACEAE

Some call this the "royal family" of plants since many of its members produce extravagantly beautiful blossoms. The flower is extremely complex. There are 3 sepals and 3 petals. The lowest petal, called the lip, differs from the other floral parts in size, form and sometimes even in color. The stamens and pistil are joined in a central column which may resemble another petal. The ovary is inferior. The parallel-veined leaves are usually alternate, with smooth margins. They sheathe the stem. The fruit is a 3-valved capsule.

Although there are 20,000 species of this family in the world, there are only two in the Santa Monicas. The flowers of both are inconspicuous in size and color but are just as complicated as the showy, exotic ones from hothouses and the tropics.

Stream Orchid

Epipactis gigantea Dougl. ex Hook.
Orchid Family. ORCHIDACEAE

Dede Gilman

The Stream Orchid is a stout, leafy-stemmed herb 1 to 4 feet high. The lower leaves are wider than the upper ones and may be up to 8 inches long. The 3 to 10 flowers, barely ½ inch long, appear on little stalks near the top. Their color has been described variously as red, brown and dull purple. There are 3 petals. The "lip" petal has erect wing-like margins and a pendulous tip. The 3 sepals are yellowish-green.

As might be expected from its common name, the Stream Orchid is found in moist places and along streams. It blooms in April and May.

This is the most commonly encountered orchid of the Pacific Coast and can occur in good-sized patches. Our mountains are not noticeably moist and you will most likely find them singly or in small scattered groups. The Indians of Northern California, where it is more abundant, made a decoction of the fleshy roots for internal use when they felt "sick all over."

Epipactis is from a Greek word *epipegnus*, which was the name for Hellebore. Hellebore was the ancient name of various plants in the genus *Helleborus*, for example the Christmas Rose. Helleborine is often used as a common name for this genus of Orchid. *Gigantea* means "gigantic" and perhaps implies there is a smaller *Epipactis* somewhere.

Rein Orchid

Habenaria unalascensis (Spreng.) Wats.
Orchid Family. ORCHIDACEAE

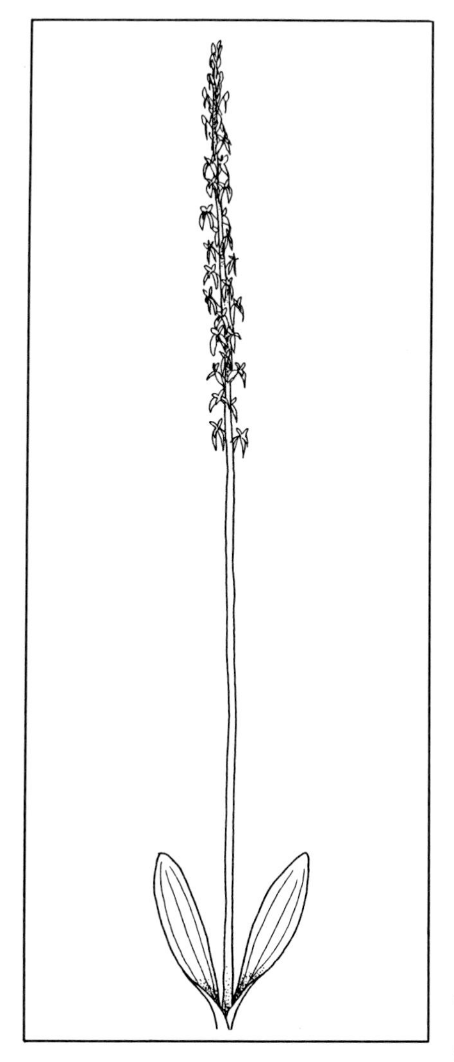

This upright plant grows 1 to 3½ feet tall with 2 to 4 narrow leaves which may be 8 inches long, although the upper ones are usually much shorter. The small greenish flowers appear in a terminal spike 4 to 12 inches long. The "lip" petal forms a spurred sac.

The Rein Orchid is relatively rare, but may be found occasionally in recently burned oak woodlands. It blooms in April and May.

If you are lucky enough to encounter this shy member of a showy family, bend close to enjoy the delicate perfume.

Habenaria in Latin is "the rein of a horse" because of the shape of its spur. The species name indicates that it was first found in the Aleutians. The Aleutians are also the westernmost locality where this species can be found.

DICOTS

Carpetweed Family. AIZOACEAE

The members of this family are usually herbs with simple, opposite, often succulent leaves. The flowers have both stamens and pistils and are symmetrical. The 4 to 8 sepals are joined. There may be no petals or many linear ones. The fruit is a capsule or is berry-like.

Most of the 1200 species of the AIZOACEAE are South African. A few have become almost weeds in America, especially those lumped under the common name of Ice Plant.

Sea Fig

Carpobrotus aequilaterus (Haw.) N.E. Brown.
Carpetweed Family. AIZOACEAE

Dede Gilman

The Sea Fig is a succulent perennial with stems that can be over 3 feet in length and forms extensive mats. The thick 3-sided leaves are not curved, have smooth margins and can be up to 4 inches long. The magenta or rose-violet flowers are 1 to 2 inches across and have numerous linear petals and many stamens.

Sea Fig is common on dunes and bluffs along the coast. It blooms from April to September.

Both species of the genus have been planted on steep slopes farther inland in the mistaken notion that they prevent erosion. Actually they are shallow-rooted and give way readily in heavy rain.

You may find *C. aequilaterus* in some floras as *Mesembryanthemum chilense.*

Carpobrotus comes from two Greek words meaning "edible fruit." The species name refers to "equal sides of the leaves."

Hottentot Fig
Carpobrotus edulis (L.) Bolus.
Carpetweed Family. AIZOACEAE

Nancy Dale

This plant is much like *C. aequilaterus* but the leaves are more curved and toothed on the lower angle. The petals are yellow, drying pink. It is a native of South Africa, but has naturalized in temperate regions throughout the world. It blooms from April through August.

C. edulis is often planted along highways and banks supposedly to control erosion. A really heavy rainfall can send many square feet tumbling to the street below. There are better choices to prevent flood damage.

A recent book on growing houseplants strongly recommends that you root a few leaves in a pot for exciting bloom when snow covers your garden. You could send some to your eastern friends.

Here again you will find this species in older floras under the genus *Mesembryanthemum.*

Edulis means "edible." Perhaps this refers to the fruit which is eaten in South Africa, but then on the other hand, the leaves are said to be a pleasant addition to salads.

Common Ice Plant

Mesembryanthemum crystallinum (L.) Rotm.
Carpetweed Family. AIZOACEAE

Linda Hardie-Scott

The Common Ice Plant is a very succulent, prostrate annual with many, tiny, glistening bubbles on the egg-shaped, reddish-tinged leaves and much-branched stems. The bubbles look like minute rain drops and disappear when you crush them. The numerous petals are white, linear, aging reddish and about ¼ inch long.

It is common along the immediate coast and blooms from March to October. It is a native of South Africa.

Ice Plant was advertised in American seed lists of 1881 as a vegetable worth serving boiled and used as a garnish.

There is another species, Slender-leaved Ice Plant *(M. nodiflorum)*, with smaller petals and more narrow leaves found at Point Mugu and Malibu Lagoon. These plants add salt to the soil after death and may prevent natives from growing on such spots.

The common name is said to have arisen because it is claimed that even on the hottest day, the leaves are cool to the touch.

Crystallinum refers to the many ice-like bubbles on the herbage. *Mesembryanthemum* was originally named *Mesembrianthemum* from *mesembria*, "mid-day," because the flowers were believed to open only in the sun, but when night-blooming species were discovered, the spelling was changed so that the name indicated a flower with its fruit in the middle (*mesos*, "middle," *embryon*, "fruit").

New Zealand Spinach

Tetragonia tetragonioides (Pallas) Kuntze.
Carpetweed Family. AIZOACEAE

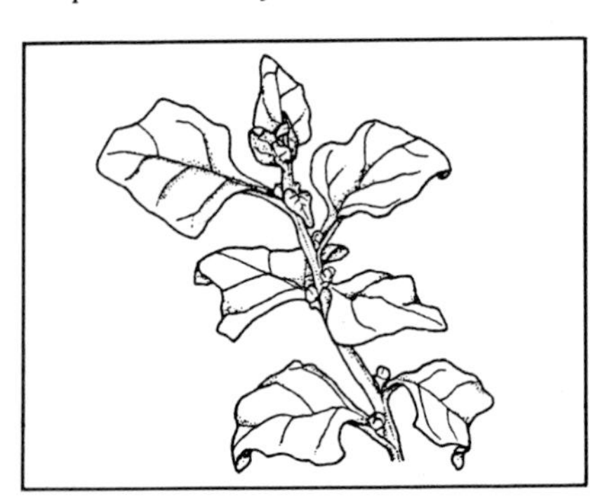

The New Zealand Spinach is a succulent, prostrate annual. This plant, like Ice Plant, is also covered with glistening little droplets filled with fluid. The large, fleshy leaves, up to 2 inches long, are triangular. The small, single flowers are greenish and the sepals number from 3 to 5. The fruit is a horned capsule.

New Zealand Spinach blooms from April to September in sandy soil and on beaches.

It is most often found in salt marshes along the coast.

As you might expect from its common name, this plant is an immigrant from the Pacific. It was first discovered when Captain Cook sailed around the world. It has been cultivated and used as a potherb in several countries and is said to be tasty in salads or steamed with lemon, butter or a sauce.

Tetra is "four" and *gonu* is "knee" in Greek, referring to the horns on the fruit. *Tetragonioides* means "like *Tetragonia.*" The species was first placed in another genus.

Sumac Family. ANACARDIACEAE

In Southern California, the members of this family are all shrubs with resinous or milky sap. The sepals and petals are commonly in 5's on a cup-like dish with 5 or 10 stamens and 3 styles. The seeds are berry-like, dry or semi-fleshy with a hard center.

It used to be easy to say that in our state we had as natives a single genus *Rhus,* with 5 species, but Poison Oak, formerly known as *Rhus diversiloba,* has now been placed in the genus *Toxicodendron.* Cashews and mangoes belong to this family.

Sumac is pronounced "shu-mac" in some parts of the United States. A learned man once pontificated that "sumac" and "sugar" were the only words in English where "su" was pronounced "shu." A listener asked, "Are you sure?"

Lemonadeberry

Rhus integrifolia (Nutt.) Benth. & Hook.
Sumac Family. ANACARDIACEAE

Steven R. Kutcher

This evergreen shrub is 3 to 10 feet high. The flat, leathery leaves, 1 to 2 inches long, are round on both ends and aromatic when crushed. The small flowers, only ⅛ inch long with 5 white to rose petals, are in compact, terminally branched clusters. There are 5 sepals and 5 stamens. The flattened, reddish fruits are covered wtih sticky bumps and short hairs.

Lemonadeberry occurs in Coastal Sage Scrub and Chaparral below 2000 feet and is more common toward the coast. It blooms from February through April.

This species is desirable as a garden ornamental. Its evergreen leaves, pinkish buds in late winter, and resistance to frost recommend it for planting singly or in masses. A cooling but rather bitter drink can be made from the fruits.

Rhus is from *rhous*, an ancient Greek name for Sumac. *Integrifolia* means that the leaf margin is not toothed.

Laurel Sumac
Rhus laurina (Nutt.) in T. & G.
Sumac Family. ANACARDIACEAE

Linda Hardie-Scott

Laurel Sumac is an evergreen shrub with smooth, reddish-brown bark that grows 6 to 15 feet tall. The oblong leaves which might be 4 inches long are untoothed on the margins, slightly folded along the midrib, have an abrupt sharp tip and smell like bitter almonds when crushed. The tiny, white flowers which are enjoyed by the bees cluster densely at the ends of the branches. The fruit is smooth and white and yields an oil.

This is a common shrub throughout our mountains and blooms in June and July.

The evergreen foliage, reddish branches and relative freedom from garden pests makes this a desirable ornamental. However, it cannot tolerate a freezing temperature.

Laurina means "laurel like."

Sugar Bush

Rhus ovata Wats.
Sumac Family. ANACARDIACEAE

Linda Hardie-Scott

Sugar Bush is an evergreen shrub 5 to 15 feet tall with thick, leathery, ovate leaves ending in a sharp point and always smooth on the margins. The leaves are up to 3 inches long and are more or less folded along the midrib. The pinkish-white flowers are in dense clusters of short spikes. The sugary, waxy, reddish fruit is covered with soft hairs.

Sugar Bush inhabits Chaparral and Oak Woodland throughout the mountains away from the immediate coast. It blooms from March to May.

R. ovata has glossy evergreen foliage and adapts to various soil and climate conditions making it more desirable for ornamental use than any of the other members of the genus. You may encounter it in gardens from San Francisco to San Diego.

The species was named *ovata* because of the ovate leaves.

Squaw Bush

Rhus trilobata Nutt. ex T. & G.
Sumac Family. ANACARDIACEAE

Dede Gilman

Squaw Bush is a deciduous shrub with stems up to 5 feet tall and slightly hairy branches. Although it is easy to confuse Squaw Bush with Poison Oak, *R. trilobata* has no allergy producing attributes. True, its leaves are divided into 3's. Sometimes the end leaf is so deeply notched it appears to be in 5's. This terminal leaflet lacks a distinct stalk which is always present in Poison Oak. The yellowish clustered flowers appear in early spring before the first leaves at the terminal end of the stem. The small (½ inch) fruits are flattened, reddish, sticky and hairy.

R. trilobata grows in canyon bottoms among Sycamores and Oaks away from the coast. It blooms in March and April.

As its common name implies, native women used the split stems for wrapping the coils of their basket material and steeped the berries in water to make a refreshing drink. It is reported that they made a lotion of the dry, powdered berries as a remedy for smallpox.

Here is a recipe for "Rhus Juice," an Indian beverage. Dry the berries of this species in the sun. Grind. Mix a cup of the result with 4 cups of water. Let stand 8 hours and sweeten to taste with honey.

Trilobata means "3-lobed."

Poison Oak

Toxicodendron diversilobum (T. & G.) Greene.
Sumac Family. ANACARDIACEAE

Dede Gilman

Dede Gilman

Poison Oak is a small shrub 2 to 6 feet high or sometimes a large vine climbing trees by adventitious roots. The leaf is divided into 3 leaflets which are rounded with small teeth. They are bright green in early spring but turn dark red in the fall. The inconspicuous, little, greenish-white flowers droop on short stems from the leaf axis. The ¼ inch fruit is white and berry-like.

Poison Oak is not a true oak but is closely related to the eastern Poison Ivy. It is one of the most widely distributed shrubs in California and is common throughout our mountains. It blooms in March and April.

This attractive plant secretes a non-volatile juice highly poisonous to many people. Severe blistering and itching results from direct skin encounter or from secondary contact on clothes, shoes or pets that have brushed against any part of the plant. Smoke may carry the oil particles, so burning as a method of elimination is hazardous. Many folk remedies are suggested and may even cure some individuals, but it is best if afflicted to consult a dermatologist.

Indians are said to have been immune and to have used the fresh juice which produces a black stain to ornament their utensils and to cure warts and ringworm. The stems were used in their baskets.

The flowers are very fragrant and contain abundant nectar. The honey gathered from them is said to be excellent.

Some earlier and recent floras will give the genus *Rhus* for this plant. *Toxicodendron* means "poison tree" and *diversilobum* means "diversely lobed."

Carrot Family. APIACEAE (UMBELLIFERAE)

This is a large family of herbs with hollow stems and alternate leaves, most often compound and usually swollen at the base. The small flowers appear in compound umbels—that is, the stalks which radiate from the top of the stem bear not single flowers but small groups of flowers again on little stems. There are 5 sepals, 5 petals, 5 stamens, and 2 styles. The ovary is inferior and the fruit is dry, ribbed or winged and is necessary to identify most of the genera and species.

This large family includes useful garden vegetables such as parsley, carrots and celery as well as dill, anise and others found in spice cabinets. We have as immigrants Sweet Fennel, Common Celery and Poison Hemlock.

There are 16 genera and 23 species in the Santa Monica Mountains, most of them are difficult to identify without a microscope and a graduate degree in botany. Many are small, inconspicuous and only rarely seen.

Common Celery

Apium graveolens L.
Carrot Family. APIACEAE (UMBELLIFERAE)

The Common Celery is a biennial with forking stems 2 to 4 feet tall. The ribbing on the stalks is distinctive. The leaves are coarsely divided; the lower ones long-stalked with 5 to 10 leaflets; the upper ones with or without stalks and usually 3 leaflets. The flowers are whitish-green appearing in a double umbel, typical of this family. The small (1/12 inch long) fruits are smooth and oval with 5 prominent ribs.

This is a common plant in wet places at low elevations throughout our mountains. It blooms from May to September.

Common Celery, a garden plant naturalized from Europe, has become a weed along streams and marshy places. The wild plant has a much stronger aftertaste than the cultivated one. It is better cooked than eaten raw.

Apium is an ancient Latin name for celery or parsley. *Graveolens* means "strong smelling."

Poison Hemlock

Conium maculatum L.
Carrot Family. APIACEAE (UMBELLIFERAE)

Barry Silver

Poison Hemlock is a tall, branching, biennial weed sometimes attaining a height of 10 feet. The large leaves are finely dissected and said by some to have a mouse-like odor. The smooth, stout stem is dotted with purple marks. The white flowers bloom in large, open, umbrella-like arrangements (umbels). The oval fruits have prominent, wavy ribs.

Poison Hemlock is widely scattered at lower elevations and flowers from May to July. It is a native of Eurasia and is now widespread throughout California and has in fact invaded all of the United States except an area between Montana and Minnesota.

All parts of the plant are poisonous, even fatal. This was the plant that provided the potion given to Socrates in ancient Greece. Look for those purple spots on the stem and avoid ingesting, or even tasting, any part of it.

Conium comes from the Greek name for hemlock. *Maculatum* meaning "spotted" tells us about the purple splotches on the stems.

Sweet Fennel
Foeniculum vulgare Mill.
Carrot Family. APIACEAE (UMBELLIFERAE)

Barry Silver

Sweet Fennel is a stout, erect perennial, 3 to 6 feet high with streaked, branching stems. The herbage is dark green and aromatic. The leaves are finely dissected into thread-like segments. The inflated leaf stalks clasp the stems. The small, yellow flowers are in double umbels. The fruit is oblong with prominent ribs.

Sweet Fennel is frequent in waste places and roadsides throughout our mountains. It is in bloom from May to September.

F. vulgare is a native of the Mediterranean region and has become quite abundant in California. It has a pronounced odor and taste similar to licorice or anise. The anise seed used commercially comes from a related cultivated plant, *Pimpinella anisum*. Licorice, however, comes from the root of *Glycyrrhiza glabra,* a member of the Pea Family.

The leaves of Sweet Fennel can be used in sauces or cooked as a vegetable. Care must be taken not to confuse Sweet Fennel with Poison Hemlock since they resemble each other when not in bloom. Sweet Fennel always has a definite licorice odor but no purple spots on the stalks. Poison Hemlock does not smell of licorice and does have the purple blotches.

It is reported that in Spanish days Sweet Fennel was spread on the mission floors to give a pleasant scent when bruised by the feet of the congregation.

Foeniculum is a diminutive of a Latin word meaning "hay" given because of the smell. *Vulgare* means that it is "common."

Sharp-toothed Sanicle
Sanicula arguta Greene ex Coult. & Rose.
Carrot Family. APIACEAE (UMBELLIFERAE)

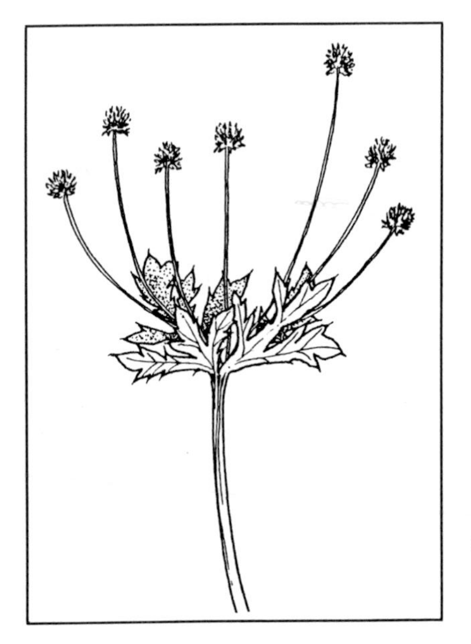

The sharp-toothed Sanicle is an upright plant ½ to 1½ feet high. The leaves are mostly basal, palmately divided into 3 to 5 parts with rough, spine-tipped teeth. There is a broad-toothed wing on the leaf stems. The flowers are bright yellow in a cluster of compact, flat-topped heads about ¼ inch across. The egg-shaped fruits are about ¼ inch long and have hooked bristles.

S. arguta is common on open, grassy slopes throughout and blooms from February to April.

We have 3 other species of *Sanicula*, 3 of *Lomatium* and 2 of *Tauschia*. The herbage of all of them probably contains alkaloids and should be considered inedible. Most of them somewhat resemble the plant described here. Unless you have both flowers and fruit and quite a bit of expertise it is exceedingly difficult to distinguish them. *S. arguta* is the most frequently encountered.

Sanicula is a diminutive of a Latin world meaning "to heal." *Arguta* means "sharp-toothed" referring to the leaves.

Milkweed Family. ASCLEPIADACEAE

The milkweeds are known for their milky juice and their odd flower structure. The petals of the 5-parted flowers are reflexed and the anthers unite to the stigma in the form of a crown with 5 hood-like appendages. Each flower can form 2 follicles from its 2 ovaries. The numerous seeds bear tufts of silky hairs at their tips for efficient wind dispersal.

In our mountains we have only the one genus and three species, two of which are seldom encountered.

The milkweed genus is the larval food-plant of the Monarch Butterfly.

Narrow-leaved Milkweed
Asclepias fascicularis Dcne. in A. DC.
Milkweed Family. ASCLEPIADACEAE

Dede Gilman

The Narrow-leaved Milkweed is an erect plant 1 to 2 feet tall with smooth herbage. The whorled, linear leaves are 2 to 5 inches long. The flowers are 5-lobed, reflexed, white to lavender. The fruit is smooth, narrow, sharp-pointed and about 3 inches long with many tufted seeds.

A. fascicularis is common in disturbed places and blooms from May to October.

Our other two milkweeds, *A. eriocarpa* and *A. californica,* are noticeably woolly and not often seen. The various species of milkweed are the larval food-plant of the large orange and black migratory Monarch Butterfly *(Danaus plexippus)* which over-winters in coastal foothill areas of California. On her travels north the female lays up to 700 eggs. One expert claims that she deposits only one egg to a plant which, if true, would mean that an abundant supply is required. Alkaloids in *Asclepias* render the Monarch unpalatable as food for birds.

The Indians had many uses for milkweeds. They found the blossoms sweet and spicy when eaten raw. They heated the milky stem juices until solid, added bear fat and made chewing gum. They utilized the tough fibers in numerous ways.

Fascicularis tells us that the "leaves are bunched." Linnaeus (1707-1778) named the genus for the Greek god of healing, *Asklepios* whom the Romans called "Aesculapius."

Sunflower Family. ASTERACEAE (COMPOSITAE)

Herbs, shrubs and even trees are in the Sunflower Family with flowers that fool the beginner. What seems to be a single flower is actually a cluster of many flowers of two kinds. The strap-shaped forms on the outer edge that look like petals are each a complete flower with all of the parts that a flower needs. They are called ray florets. The tightly packed tubular forms in the center are also complete flowers and are known as disk florets. Some members of this enormous and complex family have only rays, like the dandelions and chicory. Others, such as the thistle, have only disk florets. A great many, of course, have both. These florets are gathered together on a structure called a receptacle and underneath the receptacle is a circle of bracts known as the involucre. When you pick one blossom of the Sunflower Family, you are holding a bouquet.

If you have ever puffed a spent dandelion into the wind, you were blowing pappus, a tuft of bristles, which most species in this family have attached to the fruits in one form or another. The fruits of this family are all small and one-seeded and do not open upon maturity.

There are more than 12,000 species of ASTERACEAE around the world making it one of the largest families. So, it has been divided into several tribes. Locally we have 12 tribes, 67 genera and 119 species. Many of them are introduced weeds.

White Yarrow, Milfoil

Achillea millefolium L.
Sunflower Family. ASTERACEAE (COMPOSITAE)

Dede Gilman

White yarrow is a perennial weed 1 to 2 feet high. The strong-scented leaves are lance-shaped in outline, 2 to 4 inches long, but finely dissected into fern-like divisions. The numerous, small, flower heads are in flattened clusters at the ends of the stems. Both ray and disk florets are white. The few bracts are papery and there is no pappus.

White Yarrow blooms from May through June. This is a widespread, variable, circumpolar species with a variety of forms in California. The native varieties have apparently hybridized with an introduced one from the Old World.

The Spanish Californians used the leaves steeped in hot water for cuts and bruises and staunched blood in recent wounds with fresh branches. White Yarrow contains glucosides and alkaloids which can make sheep very sick if they eat a lot of it.

The genus was named in honor of "Achilles, Greek hero of the Trojan War," who is said to have used it to staunch the wounds of his soldiers. *Millefolium* means "a thousand leaves."

Chamisso's Burweed

Ambrosia chamissonis Less.

Sunflower Family. ASTERACEAE (COMPOSITAE)

Barry Silver

Chamisso's Burweed is a sprawling perennial forming mats on the sand 3 to 9 feet across and 6 to 12 inches high. The silvery leaves are toothed, deeply cut and covered with soft, short hairs. The male flower heads are bowl-shaped and slightly hanging from terminal spikes. The female heads are clustered below where the leaves meet the stems. They become a sharp prickly bur.

This burweed is common on dunes and sandy flats along the entire coast and serves the useful purpose of stabilizing the dunes and preventing wind erosion. It blooms from March to September.

Ambrosia is from Greek meaning "food of the gods" and is the classical name for various plants. Its application to these weedy specimens is obscure. The species name honors Adelbert von Chamisso (1781-1838), the German poet, writer and naturalist who visited the California coast in 1816 with a Russian expedition that also included Johann Friedrich Eschscholtz (1793-1831) for whom Chamisso named the California Poppy.

Western Ragweed

Ambrosia psilostachya DC. var. *californica* (Rydb.) Blake.

Sunflower Family. ASTERACEAE (COMPOSITAE)

The Western Ragweed is a coarse, perennial, aromatic herb 1 to 2 feet high. The 2 to 5 inch long alternate leaves are deeply lobed and covered with short, stiff hairs, as are the stems, which gives the plant a grayish appearance. The inconspicuous, greenish flowers are either male or female and appear on the same plant in small round heads.

It is a native of the United States and is common in all plant communities in the Santa Monicas. It blooms from August to October and is one of the principal, late summer, hay fever plants.

A. acanthicarpa and *A. confertiflora* are only occasionally found in our mountains. In older floras the genus of these two and *A. chamissonis* is given as *Franseria.*

Psilostachya is from two Greek words that mean "a bare spike."

Coastal Sagebrush

Artemisia californica Less.
Sunflower Family. ASTERACEAE (COMPOSITAE)

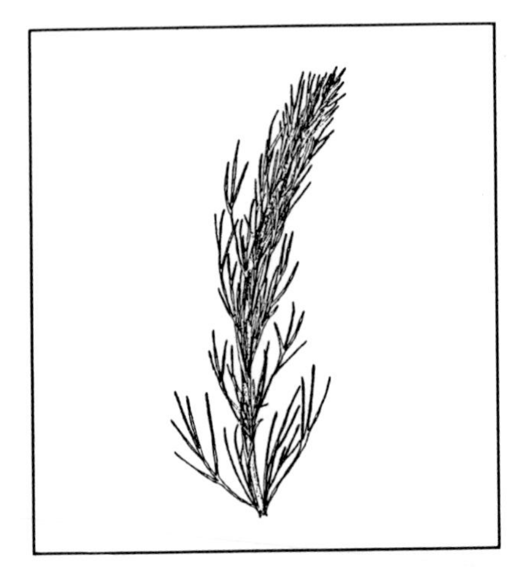

Coastal Sagebrush is a much-branched shrub 2 to 5 feet tall. The numerous, grayish-green leaves are once or twice-parted into thread-like divisions. The greenish flower heads, composed of disk florets only, are very small, nodding on tiny stalks crowded along 4 to 12 inches of the terminal stems.

This is one of the dominant plants in Coastal Sage and occurs throughout our mountains usually below 2000 feet. It blooms from August to February.

The leaves of Coastal Sagebrush have a clean, bitter, pleasantly aromatic fragrance. The Spanish Californians called the plant "Romerillo" and regarded it as a panacea for all ills. They drank a tea of it for bronchial troubles and used a strong wash of it for wounds and swellings. Early miners are reported to have spread sprays of it on their beds to drive away fleas. Even though it smells like sage and has the common name of sage, it is NOT a true sage or *Salvia*.

Tarragon *(A. dracunculus)* and Absinthe *(A. absinthium)* are cultivated for seasonings and beverages.

The Greek goddess Artemis (Diana in Roman mythology) is said to have benefitted so much from a plant of this family that she gave it her own name of Artemis.

California Mugwort

Artemisia douglasiana Bess. in Hook.
Sunflower Family. ASTERACEAE (COMPOSITAE)

California Mugwort is a perennial herb with erect stems that are woody at the base and 3 to 6 feet tall. The leaves are variable in shape, the lower ones cleft with down-pointing incisions, green above and woolly-white beneath, and 2 to 6 inches long. The small, greenish heads, composed of disk florets only, are arranged in dense, leafy spikes. The bracts are grayish and woolly.

This is a common plant of low places in Coastal Sage and Oak Woodland throughout. It blooms from July to November.

A. douglasiana is used by some to prevent or to alleviate poison oak rash. The curious common name comes from two Anglo-Saxon words, "mug" meaning "midges" (small flying insects) and "wort" meaning "herb." A relative of long ago must have either attracted the small insects or repelled them.

David Douglas (1798-1834), an ardent Scottish collector in northwestern America, collected nearly 500 specimens of California plants for the Royal Horticultural Society in England.

Mule Fat

Baccharis glutinosa Pers.
Sunflower Family. ASTERACEAE (COMPOSITAE)

Dede Gilman

Mule Fat is a large, willow-like, evergreen shrub with many, whippy, woody stems 4 to 12 feet high. The narrow, lance-shaped leaves are 1 to 4 inches long. The flesh-colored flower heads are about $^1/_3$ inch high with many disk florets and no rays. They appear in small, flattish clusters at the ends of the side branches. The papery bracts are overlapping.

Mule Fat blooms all year and is common about lakes and streams in Coastal Sage, Oak Woodland, and Chaparral throughout. On the floodplains of streams it often occurs in dense thickets.

This shrub holds its evergreen foliage through the winter and was thought to be a favorite browse for livestock.

"Bakkaris" was the name given by the ancient Greeks to a plant with a fragrant root. Stearn suggests it was named in honor of Bacchus, god of wine. Linnaeus recycled the name.

Glutinosa means "sticky" and refers to the leaves. Older floras may list *B. viminea* as a second but very similar species. It is now thought that these two are one species.

Coyote Brush
Baccharis pilularis DC. ssp. *consanguinea* (DC.) C.B. Wolf.
Sunflower Family. ASTERACEAE (COMPOSITAE)

James P. Kenney

Coyote Brush is a much-branched, evergreen shrub 3 to 12 feet high. The numerous, small leaves, less than an inch long, are egg-shaped, attached at the narrow end and have 5 to 9 coarse teeth. The dirty-white flower heads are $^{1}/_{6}$ to $^{1}/_{4}$ inch long, clustered singly at the ends of branches or in the leaf axils. Ray florets are absent. Male and female flowers are on different plants. The male ones are smaller and yellowish. The bracts are narrowly oblong and pointed at the end.

Coyote Brush is frequent near the coast and in Coastal Sage and Oak Woodland throughout. It blooms from August to November.

Sometimes this shrub is known as Chaparral Broom. A horticultural version of the subspecies *pilularis* has been cloned by Rancho Santa Ana Botanic Garden for use as a most attractive, hardy, ground cover, especially useful on banks and slopes.

Pilularis generally means "having globules," referring either to galls on the stem or the flower buds. *Consanguinea* means "related by blood."

Hairy Bur-marigold

Bidens pilosa L.
Sunflower Family. ASTERACEAE (COMPOSITAE)

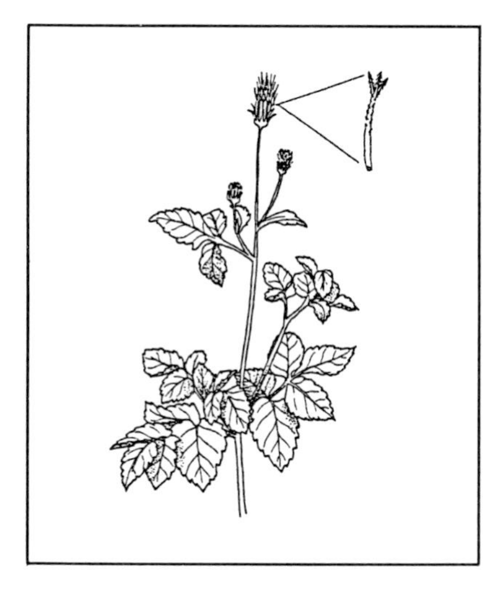

The Hairy Bur-marigold is an annual 1 to 4 feet high. The leaves have 3 to 5 egg-shaped leaflets. The margins are toothed and covered underneath with harsh hairs. The yellowish heads are inconspicuous since the ray florets are either minute or missing. The oval bracts are green in the center with thin, membranous edges. The achenes have 2 to 4 bristles topped with sharp barbs.

This native of the American tropics is now a frequent weed in lowlands throughout. It blooms from February to November.

The common name for this genus is Beggar-ticks. Not only do the achenes resemble ticks, but they stick tight like ticks, begging a ride in the hair of animals or hooking into boots, jeans or socks and thus travel widely to new locations.

We have two other species in our mountains. *B. frondosa* with smoother leaves is located near Lake Sherwood. *B. laevis* with simple leaves if found, but rarely, near the Los Angeles River.

Bidens is from Latin meaning "2-toothed" and refers to the bristles on the achenes. *Pilosa* means "hairy."

California Brickelbush

Brickellia californica (T. & G.) Gray.
Sunflower Family. ASTERACEAE (COMPOSITAE)

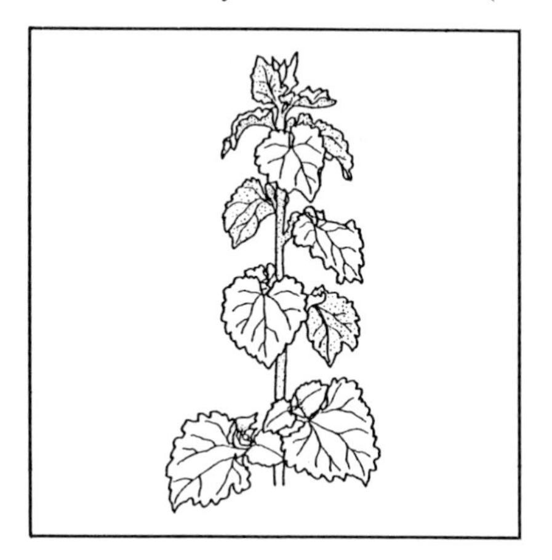

California Brickelbush is a spreading shrub up to about 3 feet in height with broad, rounded leaves on short stalks. The leaves are ½ to 2 inches long with minute teeth around their margins, covered with short, gray hairs and heart-shaped at the base. The creamy heads are about ½ inch long, all disk, in small terminal clusters on lateral branches.

California Brickelbush is common on dry slopes in Coastal Sage and Chaparral throughout. It blooms from August to November.

The flowers of *B. californica* exude a very pleasant odor in the evening. We have only one other species, *B. nevinii,* which is more spreading in habit, less common and has larger flower heads. There are 14 native species in the state. Where they grow near each other they are thought to hybridize. This genus is known to be native only in the Americas.

The genus is named in honor of John Brickell (1749-1809), an early naturalist and physician of Georgia who came to the United States in 1770 from Ireland and settled in Savannah in 1779.

Yellow Star Thistle, Tocalote

Centaurea melitensis L.
Sunflower Family. ASTERACEAE (COMPOSITAE)

Geoffrey Burleigh

The Yellow Star Thistle is an erect, much-branched annual 1 to 2 feet high. The narrow, lance-shaped leaves are alternate, gray-green and have edges that run down the stems. The flower heads are small in comparison to other thistles and often exhibit cottony hairs. The yellow florets are all tubular and rise out of a globular structure armed with prickly spines.

This immigrant from southern Europe is common in fields and by roadsides throughout. Although its cousins, the Cornflowers and Dusty Millers are favored in the garden, the Star Thistle, a common impurity in seed grain, has become particularly obnoxious in agricultural fiields and pastured hills. It blooms from May through July.

The popular name Star Thistle was given because the spiny heads suggested a medieval weapon—a metal ball on a long handle set with spikes called a "morning star."

Centaurea is based on the ancient Greek name for "Centaur (kentauros), a mythical creature, half man and half horse." Centaury was a plant in Greece with medicinal properties that were said to have been discovered by the Centaur, Chiron. The species name, *melitensis,* means "from Malta."

White Chaenactis

Chaenactis artemisiaefolia (Harv. & Gray) Gray.
Sunflower Family. ASTERACEAE (COMPOSITAE)

Margaret Stassforth

The White Chaenactis is a stout, branching annual 1 to 5 feet high. The alternate leaves are about 4 inches long, divided 2 or 3 times and are sticky-hairy. The white flower heads, all disk florets, are on leafless, branched stalks. The green, cup-like structure beneath the flower head is sticky to the touch.

White Chaenactis blooms from April to June in open places in Chaparral from Triunfo Pass east to Sepulveda Canyon. It is most abundant after fires.

Another white-flowered species, Fremont Pincushion *(Chaenactis fremontii)*, is common on the Colorado Desert. The marginal florets of this one are much enlarged, the plant is more slender and seldom ever a foot tall.

Chaenactis means "a gaping ray" and was given because in most species the outer florets are enlarged into a wide open mouth. (Pronunciation of the generic name is not difficult—"ke-NAK-tis.") *Artemisiaefolia* mean that the "leaves resemble those of artemisia."

Yellow Chaenactis

Chaenactis glabriuscula DC.
Sunflower Family. ASTERACEAE (COMPOSITAE)

Margaret Stassforth

The Yellow Chaenactis is an erect annual 6 inches to 1½ feet high. The thickish leaves are once or twice divided into a few, very narrow lobes. The flower heads are golden-yellow and up to 1½ inches across. The florets are all disk. The outer ones flare like a top and are larger than the inner ones giving the false effect of rays about disks.

Yellow Chaenactis can be found on sandy flats just behind the coast, most particularly at Point Dume. It is encountered on open slopes in Chaparral elsewhere. It blooms from April to June.

The neat, roundish-topped heads of this flower are quite suggestive of the common name "Pincushion" given most accurately to the genus. This name is especially appropriate when the flower head is studded with the protruding stamens and pistils.

Glabriuscula is composed of two Latin words meaning "smooth" and "little."

Western Thistle, Cobweb Thistle
Cirsium occidentale (Nutt.) Jeps.
Sunflower Family. ASTERACEAE (COMPOSITAE)

Dede Gilman

The Western Thistle is a stout biennial 2 to 5 feet high. The stems are very white when young. The rosette leaves which form the first season are lance-shaped and slightly lobed with prickly edges. The stem leaves have sharp teeth ending in spines. The whole plant is festooned with cottony hairs. The disk florets are purplish-red in a closely packed head. There are no rays.

Western Thistle blooms from March to July in Coastal Sage and Chaparral.

C. californicum is very similar but the flowers are distinctly pink. It does not begin to bloom until April. Bull Thistle *(C. vulgare)* has purple flowers and, unlike the other two, is not a native, but came from Eurasia. It does not begin to bloom until June, but continues until October.

Cirsium is from Greek, *kivsion,* "a kind of thistle." *Occidentale* means "west."

Bigelow Coreopsis
Coreopsis bigelovii (Gray) Hall.
Sunflower Family. ASTERACEAE (COMPOSITAE)

Geoffrey Burleigh

This is an erect annual with smooth herbage, 4 to 24 inches high. The narrowly dissected leaves are mostly at the base. The flower heads, both ray and disk florets, are golden-yellow and showy. They appear singly at the end of a leafless stem.

Bigelow Coreopsis has previously been reported as rare in Grassland and Oak Woodland from Malibu Creek eastward at low elevations away from the coast but has recently been encountered much more frequently. It blooms in March and April.

This plant is more common on the deserts than in our mountains. In the northern deserts it often produces spectacular sheets of color. *C. douglasii* with smaller flowers which you may encounter elsewhere in coastal areas has not established itself locally.

Coreopsis means "bug-like" from the achenes of the original species. The species name honors Dr. J. M. Bigelow (1804-1878), a professor of botany and pharmacology at Detroit Medical College, compiled a flora of Lancaster, Ohio, in 1841. He collected in the West under Whipple (See *Yucca whipplei*) in the Pacific Railroad survey of 1853. Many other California plants bear his name.

Giant Coreopsis, Sea Dahlia
Coreopsis gigantea (Kell.) Hall.
Sunflower Family. ASTERACEAE (COMPOSITAE)

Linda Hardie-Scott

The Giant Coreopsis is a stout shrub that can be up to 10 feet tall with a tree-like trunk 5 inches thick. The light green leaves are divided into very thin fine parts and appear in spring in tufts at the top of the chunky trunk. The flower heads are bright yellow, large (up to 3 inches across) and showy at the top of naked 6-inch flower stalks.

Giant Coreopsis blooms from March through May on the immediate coast from Zuma Beach west. Another colony appears inland near Camarillo State Hospital. Outside of our area this spectacular native is found primarily on the Channel Islands with 1 or 2 rare showings along the coast south of San Luis Obispo.

Until blooming time, the ugly, brown trunk and dead foliage give no hint of the glory to come. The resemblance of this flower to a single yellow dahlia has suggested one of its common names.

Gigantea means, of course, "very large" and the overall plant is that.

Cudweed-aster, Common Corethrogyne
Corethrogyne filaginifolia (H. & A.) Nutt. var. *virgata* (Benth.) Gray.
Sunflower Family. ASTERACEAE (COMPOSITAE)

Geoffrey Burleigh

The Cudweed-aster is a slender perennial up to 3 feet high. The lance-shaped leaves can be over 2 inches long. The ¼ inch flower heads have violet rays and purple disks. The bracts below overlap and their green tips recurve at maturity. The stems and leaves are covered with white wool.

This is a common plant of late summer in the dry, brushy chaparral, blooming from July to November.

This is the only *Corethrogyne* in the Santa Monicas, but other species with numerous varieties occur throughout California. We have only one other variety on the coastal bluffs in the Santa Monicas, *C. filaginifolia* var. *latifolia,* which has bracts covered with matted, white hairs at the time of flowering.

Corethrogyne is from the Greek meaning "broom" and "female" referring to the brush-like style tips. The species name refers to the white, wool threads on the leaves resembling those on the genus *Filago,* another member of the Aster Family.

Brass Buttons
Cotula coronopifolia L.
Sunflower Family. ASTERACEAE (COMPOSITAE)

James P. Kenney

Brass Buttons are herbaceous, rather succulent, strong-scented perennials, a foot or so high, usually with many clustered stems spreading on the ground. The yellowish-green leaves, alternating and clasping the stem, assume many shapes and forms on the same plant. The flower heads are roundish-flattened, about ⅓ inch across and borne on long, naked stalks. Ray florets are absent but there are numerous, bright yellow, disk florets. There are 2 ranks of scaly bracts.

Brass Buttons bloom from March to December and are common along streams and the banks of Salt Marshes.

They are native to South Africa but came early to our state. The Spanish Californians called the plant *"Boton de Oro,"* that is, Gold Button. The foliage when crushed gives out an odor between lemon-verbena and camphor. They make a lively sparkle of color in wet places along the coast.

Cotula is from Greek meaning "a small cup" and refers to a hollow at the base of the leaves. *Coronopifolia* means "the leaves are like those of *Coronopus.*" *Coronopus*, whose common name is Swinecress, is a naturalized European member of the Mustard Family. *Coronopus* is from two Greek words meaning "crown" and "foot."

California Encelia, Bush Sunflower
Encelia californica Nutt.
Sunflower Family. ASTERACEAE (COMPOSITAE)

Stanley J. Higgins

The Bush Sunflower is a bushy perennial up to 5 feet high. The alternate leaves grow close to the main stem, broadly lance-shaped, about 2 inches long, green on both sides and prominently 3-veined from the base. The flower heads are showy with golden-yellow rays and brownish-purple disks. They grow singly on long stalks.

E. californica blooms from March to June in Coastal Sage and Chaparral throughout, spreading into disturbed areas below 2000 ft.

This plant has a strong odor and is rough to the touch, but it is very colorful and attractive. A close cousin that does not grow in our mountains, Incienso or Brittlebush *(E. farinosa),* is abundant east of us and into the deserts. The foliage of Incienso is silvery-white with a dense, scurfy wool, and it exudes bubbles of resinous gum sometimes used in the past for incense, chewing gum, varnish and pain killers.

Christoph Entzelt, an early Lutheran clergyman, who Latinized his name to Encelius, wrote a book in 1551 about the medicinal uses of minerals, animal parts and plants. His book is quoted as a source of plant names.

Leafy Daisy, Fleabane
Erigeron foliosus Nutt. var. *foliosus.*
Sunflower Family. ASTERACEAE (COMPOSITAE)

James P. Kenney

The Leafy Daisy is a tall, branched, smooth perennial 8 to 40 inches high. The narrow leaves are ¾ to 2¼ inches long. The numerous flower heads with 20 to 60 narrow lavender rays and yellow disk florets appear openly at the top.

Leafy Daisy is a common plant of the Chaparral and blooms from May to July.

There are many species of Fleabane throughout the state, but we only have this and one other variety (var. *stenophyllus*) which has narrower leaves and is not as common.

Many years ago it was believed that the plants of this genus were harmful to fleas so they were called Fleabane.

Erigeron is a Greek name meaning "old man in the spring" which may refer to the early flowering and fruiting of many species. *Foliosus* means "leafy."

Golden Yarrow
Eriophyllum confertiflorum (DC.) Gray.
Sunflower Family. ASTERACEAE (COMPOSITAE)

Barry Silver

The Golden Yarrow is a low shrub a foot or so high, which is white at first with a close woolliness that later disappears. The alternate leaves are deeply 3 to 5-lobed. The flower heads are golden-yellow, about ⅜ inch across with both ray and disk florets in crowded, flat-topped clusters on short stalks.

Golden Yarrow blooms from January through July and is common in open places away from the immediate coast. Its bright color and tidy growth habit make it especially noticeable.

The blossoms in form and arrangement suggest the true yarrow, *Achillea millefolium,* which is white, and previously described. Since it blooms from the first of the year well into summer it can be an addition to the garden, growing from seeds or small plants. It needs full sun and well-drained soil. Cut it back to within a few inches of the ground in fall.

Eriophyllum means "woolly leaf" referring to the matted, white hairs that cover the plant when it is young. *Confertiflorum* means that the flowers are crowded.

California Everlasting
Gnaphalium californicum DC.
Sunflower Family. ASTERACEAE (COMPOSITAE)

Stanley J. Higgins

California Everlasting is a sturdy biennial ½ to 3 feet high with green, sticky foliage. The lance-shaped leaves are green on both sides and may be 4 inches long at the base of the plant, shorter above. The numerous, rounded flower heads are clustered at the top of the stems. The white, papery, scaly bracts are much more noticeable than the minute flowers.

There are 9 species of this genus growing in the Santa Monicas and they are known interchangeably by such names as Everlasting, Cudweed and Lady's Tobacco. If the heads are clustered at the tips of the stems and the leaves surmount them the plant may be *G. palustre*. If the leaves are densely woolly underneath you may have encountered *G. leucocephalum* or *G. bicolor*. Green leaves top and bottom without wool indicate *G. californicum* (pictured here) or *G. ramosissimum*. The first has wider leaves than the second and the second has pinker flowers. If the leaves are woolly on both surfaces, the plants can be *G. luteo-album, G. chilense, G. microcephalum* or *G. beneolens*. The last two are perennials. There are other differences in these confusing species which can be checked in more technical floras.

California Everlasting is common in Coastal Sage and in open places in Chaparral throughout. It blooms from January through July. It is reported that sleeping on a pillow made of these flowers will cure catarrh, an inflammation of the mucous membranes.

The generic name is from Greek and means "a lock of wool." Most of these plants are quite woolly.

Big Gum-plant
Grindelia robusta Nutt.
Sunflower Family. ASTERACEAE (COMPOSITAE)

James P. Kenney

The Big Gum-plant is a coarse, bushy plant, woody at the base, 1½ to 4 feet high. The toothed leaves are thick, narrow, 3 or 4 inches long and clasp the stems. The yellow flower heads can be 2 inches across with both disk florets and 25 to 45 rays. They bloom singly at the tips of the stout stems. The buds are covered with a whitish gum which protects them in harsh sunlight. The green bracts underneath the heads are overlapping and have tips that curve down.

Big Gum-plant can be found blooming from April through October in dry locations throughout.

The resinous tops of this and other species of Gum-plants were used by the Indians as a wash for poison oak rash. You can try this but you will probably be better off with something prescribed by a dermatologist. (Some authorities claim that the original Indians of California were immune to poison oak.) A book published in 1922 informs us that each summer the top 5 inches of this plant were cut off and shipped by tons to the east, to be sent back later in the form of a medicine called Grindelia and used to soothe whooping cough, asthma, bronchitis and kindred complaints.

David Hieronymous Grindel (1776-1836) was a highly respected Estonian pharmacologist, physician and botanist. It is hardly necessary to mention that the plant is indeed robust.

Common Hazardia, Goldenbush

Haplopappus squarrosus H. & A. ssp. *grindelioides* (DC.) Keck.
Sunflower Family. ASTERACEAE (COMPOSITAE)

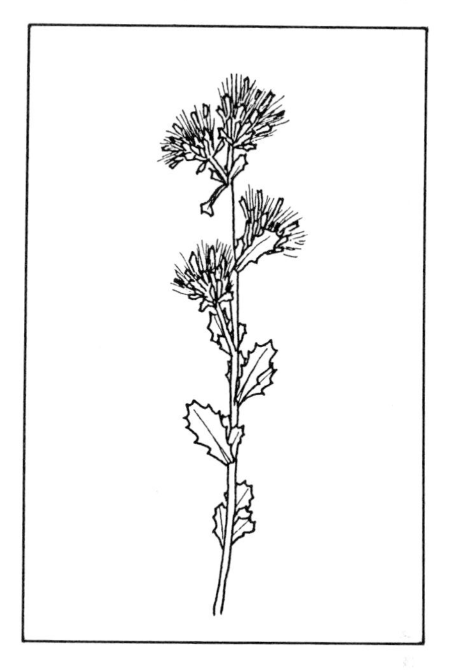

Common Hazardia is a low-branched subshrub, 1 to 3 feet tall, with brittle branchlets and numerous, small, oval, clasping leaves, that are sharply-toothed with spiny tips. The yellow florets are all disk florets and are up to ½ inch high. The many green-tipped bracts overlap like shingles on a roof.

This plant is common in Coastal Sage and Chaparral nearly throughout except along the immediate coast, blooming from August to November.

Twenty species of shrubs of this genus are native to California. We have five of them in our mountains. The leaves of the Coastal Isocoma *(H. venetus,* ssp. *vernonioides)* do not have the spiny tips of *H. squarrosus.*

Many people call *Haplopappus,* Goldenbush, and since most of them have heads of golden-yellow flowers, this name seems appropriate.

Haplopappus is a Greek name meaning "simple down or fluff" in reference to the single pappus ring. *Squarrosus* means "scaly" or "rough." The common name Hazardia goes back to a former genus name of these plants.

Common Sunflower
Helianthus annuus L.
Sunflower Family. ASTERACEAE (COMPOSITAE)

Dede Gilman

The Common Sunflower is a tall, stout, openly-branched annual, 6 feet or taller. The alternate leaves may be up to 10 inches long, oval in shape, with numerous, stiff hairs creating a sandpapery texture. The sticky flower heads are 3 to 5 inches across with yellow rays and purple-brown disk florets. The flowers appear singly at the branch ends.

Common Sunflower is a plentiful plant of fields and roadsides, blooming from February until October.

Another species, Slender Sunflower *(H. gracilentus)* is very similar but not as robust and has yellow disk florets. The Giant Sunflower of gardens is a relative of this native.

The nutritious seeds were tasty to the Native American populations. A useful fiber was taken from the coarse stalks and a good dye was made from the flowers.

In Greek, *helios* means "sun" and *anthos*, "flower." The name was given to the flower because it was supposed to turn always to the sun. Scientists explained this with the hypothesis that the stem grows more quickly on the shaded side than on the other. It worried some of the early herbalists when they encountered a variety with several blooms all facing in different directions. *Annuus* means "annual."

Santa Susana Tarweed
Hemizonia minthornii Jeps.
Sunflower Family. ASTERACEAE (COMPOSITAE)

Linda Hardie-Scott

Santa Susana Tarweed is a shrub up to 40 inches tall with as many as 500 woody stems from the base. The herbage is fragrantly resinous. The alternate leaves are slightly thickened. There are 8 ray florets and 18 to 23 sterile disk ones. The anthers are yellow, an easy distinction between this species and the much more abundant *H. ramosissima.*

One other species, *H. pungens,* is reported from the Santa Monicas, but it is introduced and almost never encountered. The Santa Susana Tarweed was for many years known only from sandstone outcrops in the Santa Susana Pass area. It was first reported in our mountains from east of Castro Peak in 1977 and has since been found near Chatsworth Reservoir, on the southwest slope of Calabasas Peak and in Charmlee Park.

The name *Hemizonia* means "half-girdle" and is applied to the genus because the bracts half encircle the ray seeds. Theodore Wilson Minthorn (1886-1967) with his sister, Maud Aileen Minthorn, collected in the Santa Susana Mountains from 1905 to 1923. He received a Masters in Botany from UC Berkeley in 1921.

Slender Tarweed

Hemizonia ramosissima Benth.

Sunflower Family. ASTERACEAE (COMPOSITAE)

Geoffrey Burleigh

Slender Tarweed is an annual herb up to 40 inches tall and widely-branched toward the top. The lower leaves are mostly gone at the time of flowering. The upper leaves are small and very narrow. The numerous flower heads are clustered at the ends of the short branches. There are 5 yellow, 3-lobed ray florets and 6 disk florets. The anthers are black. The bracts underneath are leaf-like and barely overlap.

Of the numerous plants on the Pacific Coast known as tarweeds from their disagreeable sticky exudations on stem and leaf, this is one of the most common. The viscid tar may stain your clothing and skin. Alcohol can remove it. Whole fields and hillsides appear yellow with this abundant native from May through September.

Ramosissima means "much-branched."

Telegraph Weed
Heterotheca grandiflora Nutt.
Sunflower Family. ASTERACEAE (COMPOSITAE)

Linda Hardie-Scott

Telegraph Weed is a tall, hairy, annual herb with 1 to several erect stems, 2 to 6 feet tall. The many alternate, gray-green leaves are oval, 1 to 3 inches long, slightly-toothed, hairy on both sides, and droop at night. The numerous, flat-topped, golden-yellow heads are about an inch across, subtended by many, narrow, green bracts. There are around 30 ray florets surrounding the many disk florets. Ray fruits are usually bare, whereas the disk fruits are topped by brick red straight hairs.

Telegraph Weed blooms the year round especially along roads and in waste places throughout. When you see the less than spectacular blossom, you may wonder why *grandiflora* was chosen as a species name. When asked about this a learned botanist once replied, "You should see the other one." There is only one other in California, Camphor Weed *(H. subaxillaris)* and the flowers ARE smaller.

The common name was probably given because of the tall, straight stems, like a telegraph pole. There is also a strong creosote smell about the plant that is reminiscent of telegraph poles.

Heterotheca is Greek meaning "different" and "ovary," from the unlike fruits of the ray and disk florets.

Smooth Cat's Ear
Hypochoeris glabra L.
Sunflower Family. ASTERACEAE (COMPOSITAE)

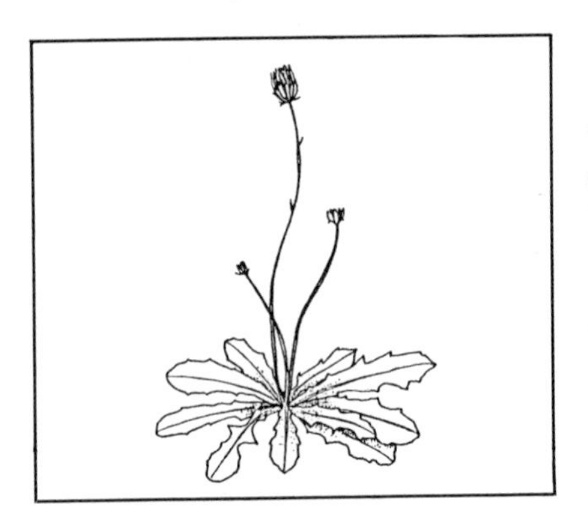

The Smooth Cat's Ear is an annual with small, flowering heads not more than ¼ inch broad with only yellow ray florets. The mostly basal, smooth leaves are lobed and up to 4 inches long. The achenes have feathery bristles ¾ inch long.

H. glabra blooms from March to June in fields and grasslands.

Smooth Cat's Ear comes to us from Europe, Western Asia and North Africa. It has become widely distributed in the state. Another species, Hairy Cat's Ear *(H. radiata)*, blooms later and is found mostly in lawns. It is also an immigrant but is a perennial, and the leaves are covered on both surfaces with yellow hairs.

The genus name is a Greek one used by Theophrastus (see *Sisyrinchium bellum*) for this or some other genus. *Glabra* means "smooth."

Common Goldfields
Lasthenia chrysostoma (F. & M.) Greene.
Sunflower Family. ASTERACEAE (COMPOSITAE)

Dede Gilman

Common Goldfields are slender annuals 6 inches or so high. The opposite leaves are needle-like, an inch or so long. The yellow flower heads at the top of the stems have both disk florets and 10 to 14 ray florets ¼ to ⅜ inch long.

Goldfields bloom in the Santa Monicas in March and April in shallow soil throughout. *L. coronaria* is very similar but not nearly as common, and its leaves are lobed.

Goldfields is a most descriptive common name since the abundance of this plant often covers fields and low hills with a close carpet of rich gold. Mary E. Parsons tells us of the fascinating past of this small plant in her charming book, *The Wildflowers of California,* first published in 1897 and still in print. In some localities Goldfields were frequented by a small fly feeding on the pollen and they became known as Fly Flowers. On early Spanish California playing cards, the Jack of Spades always held one of these flowers in his hand. The señoritas called it *si me quieres, no me quieres,*—"Love me, love me not"—a game other young ladies have been known to play with daisies.

Lasthenia remembers a Greek girl who, in order to attend the lectures of Plato, dressed in the clothes of a man. *Chrysostoma* means "goldenmouthed."

Tidy Tips
Layia platyglossa (F. & M.) Gray.
Sunflower Family. ASTERACEAE (COMPOSITAE)

Dede Gilman

Tidy Tips is a simple or branching, hairy annual 4 to 12 inches high. The toothed, hairy leaves are alternate, narrow and without stalks. Some of the upper ones are deeply cut. The flower heads at the top of the stems are about an inch in diameter. The ray florets are yellow tipped with white. The disk florets are yellow with black anthers. This is a pleasing color scheme, and it is indeed a neat and tidy flower.

Tidy Tips bloom from March to May on sandy coastal flats, particularly at Point Dume and Latigo Canyon.

The flowers are delicately fragrant. Countless millions of this little wildling once showed themselves every spring in the San Fernando Valley, but they have lost out to parking lots and freeways.

George Tradescant Lay (?1797-1845) was a botanist with Captain Beechey on the *Blossom* which visited California in 1827. *Platyglossa* means "broad tongued," probably referring to the ray florets.

Cliff Aster
Malacothrix saxatilis (Nutt.) T. & G.
Sunflower Family. ASTERACEAE (COMPOSITAE)

Linda Hardie-Scott

Cliff Aster is a tall leafy perennial, slightly woody at the base, 1 to 3 feet high and much-branched. The leaves are narrow and several times as long as wide, 1 to 4 inches long. The stems become leafless as they approach the flower heads. The white flower heads are at the top of the branches. They have ray florets only and change to pale rose or lilac in age.

Cliff Aster blooms the year round and is common in all plant communities. It likes open and disturbed areas.

The only other species in this genus in our mountains, *M. clevelandii,* is an annual with yellow flowers and is rarely seen.

Malacothrix means "soft hair," an allusion to the woolliness of the young plant. *Saxatilis* means it is "found among rocks."

California Chicory
Rafinesquia californica Nutt.
Sunflower Family. ASTERACEAE (COMPOSITAE)

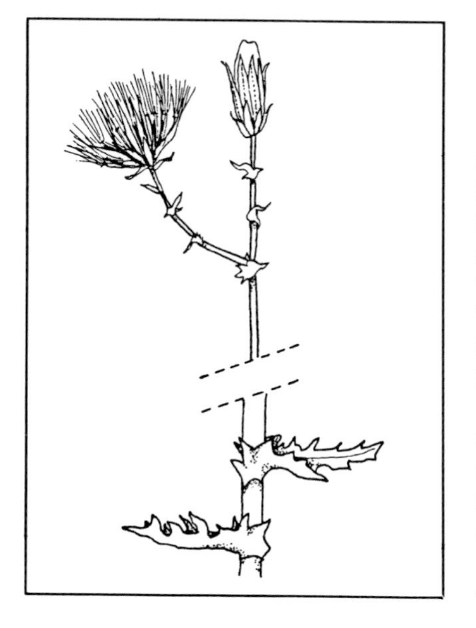

California Chicory is an annual 8 to 60 inches tall. The leaves are 2 to 8 inches long. Some are smooth-margined and some are divided. The lower leaves are on little stems while the upper ones clasp the main stem. The flower heads have 15 to 30 white ray florets only. The bracts under the flower heads are brownish bristles.

California Chicory blooms from April to June in shady places but is found only occasionally on the western end of the mountains. It is more abundant in burned over areas.

A close relative, *R. neo-mexicana,* is common on the desert and distinguished by rose-colored veins on the underside of the rays. The plant most often known as Chicory is *Cichorium intybus* with large lovely blue flower heads. It is an invader from England and is infrequent but may be encountered around Santa Monica and West Los Angeles.

The genus name honors an eccentric 19th century botanist and friend of Audubon, Constantine Samuel Rafinesque-Schmaltz (1783-1840):

> "Dedicated to the memory of an almost insane enthusiast in natural history; sometimes an accurate observer, but whose unfortunate monomania was that of giving innumerable names to all objects of nature and particularly to plants."
>
> Thomas Nuttall (1786-1859)

Shrubby Butterweed
Senecio douglasii DC. var. *douglasii.*
Sunflower Family. ASTERACEAE (COMPOSITAE)

Dede Gilman

Shrubby Butterweed is a perennial, forming a bush 3 to 4 feet high. The alternate, white-woolly leaves are divided into narrow, almost thread-like lobes. The flower heads of both ray and disk florets are yellow. There are about a dozen rays, narrow and barely ½ inch long. The heads appear in terminal loose-branching clusters. They have conspicuous, narrow, little bracts at their base.

Shrubby Butterweed is common throughout away from the immediate coast, mostly in Coastal Sage, blooming from June to November.

The genus *Senecio* comprises perhaps 1000 species distributed worldwide. More than 40 of these are indigenous to the Pacific Slope. When most of the wildflowers in Southern California have ceased to bloom, it is pleasant to encounter the glowing yellow of the Shrubby Butterweed.

Senecio is from Latin, *senex,* which means "old man" because the soft bristles on the achenes are snowy white. (See *Artemisia douglasiana* for an explanation of Douglas.)

Milk Thistle

Silybum marianum (L.) Gaertn.
Sunflower Family. ASTERACEAE (COMPOSITAE)

Barry Silver

Milk Thistle is an erect, much-branched annual or biennial 3 to 6 feet high. The large, prickly leaves are deeply lobed and wavy-margined. Their shiny, green surfaces are prominently blotched with white. The flower heads, up to 2 inches across with tubular, rose-purple florets, are solitary at the branch ends.

Milk Thistle is common in weedy places at low elevations and blooms from May to July.

The Milk Thistle came to us from the Mediterranean region but has long since worn out its welcome. It has become a persistent nuisance in cultivated grounds and crowds out less vigorous natives in many places. It also flavors milk when eaten by cows.

Once in the Old World it was thought that the white marks on the leaves resulted from drops of milk that fell from the Virgin's breast as she nursed the infant Jesus, hence the species name, *marianum*, meaning "of Mary."

Silybum is from the Greek name for a thistle that was used for food.

California Goldenrod

Solidago californica Nutt.
Sunflower Family. ASTERACEAE (COMPOSITAE)

California Goldenrod is an erect perennial herb 2 to 4 feet high. The many, oval, toothed leaves are pointed at the tip and taper to a short stalk. They have one unbranched vein. The 7 to 12 yellow, ray florets and an unequal number of disk ones are grouped in densely clustered, small heads forming a narrow, compact, compound cluster. The achenes are hairy.

Our state is not as rich in Goldenrods as New England. Neither of the two species in the Santa Monicas is found in abundance.

California Goldenrod is scattered on dry hillsides often under oaks throughout and blooms from July to October.

We also have Western Goldenrod *(S. occidentalis).* It likes wet places and blooms from August to November. Western Goldenrod has narrower leaves which are 3 to 5-veined and looser flower clusters usually with more ray florets than disk ones. Its blossoms are said to smell like Acacia.

Solidago is a Latin name meaning "to make whole." These plants were reputed to have medicinal value.

Twiggy Wreath Plant

Stephanomeria virgata Benth.
Sunflower Family. ASTERACEAE (COMPOSITAE)

James P. Kenney

Twiggy Wreath Plant is a smooth-stemmed annual, upright with rigid, pole-like stems, usually 2 to 6 feet high, but sometimes even 15 feet high. The leaves on the lower stem are wavy-toothed or deeply divided. The upper leaves are small and narrow and are mostly gone by flowering time. The pale, pinkish-lavender flower heads, purplish on the back, are an inch or less across. The flower head is all ray florets.

S. virgata is found in open disturbed places throughout and blooms from July through November.

We also have *S. cichoriacea,* a perennial with white-woolly leaves. We can be grateful that both prefer a bonedry world, blooming long after the rainy season is past when most other flowers have set their seeds and disappeared.

Stephanomeria comes from two Greek words meaning "wreath" and "division." (It is unclear why this was chosen.) *Virgata* means "wand-like" and that is applied to the tall, bare stems.

Salsify, Oyster Plant
Tragopogon porrifolius L.
Sunflower Family. ASTERACEAE (COMPOSITAE)

Geoffrey Burleigh

Salsify is a stout, smooth, somewhat succulent, unbranched, biennial herb 2 to 4 feet high. The numerous, narrow, smooth, grass-like, green leaves are 8 to 12 inches long and clasping at the base. The stem is swollen and hollow just below the single, large head of purple, ray florets. The head is 2 to 4 inches broad, and usually is open in the early morning and closed by mid-day. There is a single series of pointed, green bracts. The achenes have long, feathery bristles so the plant is showy even when the deep purple petals are gone.

Although nowhere widely distributed, this native of the Mediterranean region may be encountered on waste ground near the coast from Point Dume to Santa Monica. It blooms from April to June.

A yellow-flowered one, *T. pratensis*, is encountered occasionally also.

The thick, white taproot of Salsify is cooked as a vegetable and has a flavor similar to that of oysters. To enjoy it, you should boil it until tender, remove the outer skin, dice and season with salt and pepper.

Tragopogon is from two Greek words meaning "goat" and "beard." When it is in seed, prominent feathery hairs suggest such a beard. *Porrifolius* means that the "leaves look like those of leeks."

Canyon Sunflower
Venegasia carpesioides DC.
Sunflower Family. ASTERACEAE (COMPOSITAE)

Dede Gilman

This tall, branched perennial herb is slightly woody at the base. The thin leaves are heart-shaped, 2 to 6 inches long with rounded teeth on the margins. The yellow flower heads, with both rays and disks, are large, up to 2 inches across, appearing at the ends of the stems on short stalks.

Canyon Sunflower blooms from March to June and is frequently found in shady canyons and on rocky banks of streams below 3000 feet.

There is a striking contrast between the handsome, pale green leaves and the smooth, dark stems, and the flowers which are a particularly attractive shade of bright yellow. To offset all this loveliness, however, it does have a rather unpleasant odor.

The name *Venegasia* commemorates Padre Miguel Venegas, a Jesuit scholar and historian at a seminary at Puebla, Mexico.

Carpesioides means "like *Carpesium*," a plant in the everlasting tribe of the ASTERACEAE which grows wild from the Pyrenees to Japan. The flower heads of *Venegasia* resemble the buds of *Carpesium*.

Spiny Clotbur, Cocklebur
Xanthium spinosum L.
Sunflower Family. ASTERACEAE (COMPOSITAE)

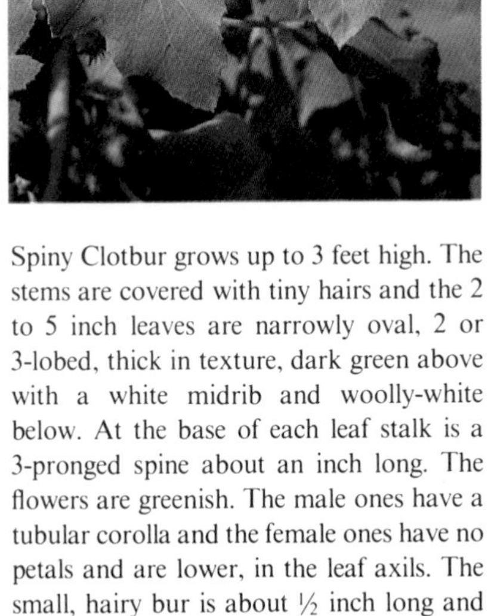
James P. Kenney

Spiny Clotbur grows up to 3 feet high. The stems are covered with tiny hairs and the 2 to 5 inch leaves are narrowly oval, 2 or 3-lobed, thick in texture, dark green above with a white midrib and woolly-white below. At the base of each leaf stalk is a 3-pronged spine about an inch long. The flowers are greenish. The male ones have a tubular corolla and the female ones have no petals and are lower, in the leaf axils. The small, hairy bur is about ½ inch long and covered with hooked prickles.

Spiny Clotbur is a common weed in fields and waste places blooming from July to October. It is not considered a native although its place of origin is unknown. *X. strumaritum*, also a common weed, lacks the 3-forked spines.

The two seeds protected in each bur retain their vitality for many years. Both species are obnoxious in pastures where the burs become tangled in manes and tails and reduce the value of wool on sheep. Both species are poisonous to animals in the seedling stage as are the seeds.

Xanthium comes from a Greek word meaning "yellow," and was the ancient name of a plant used to dye hair. *Spinosum* is Latin for "thorny."

Borage Family. BORAGINACEAE

The Borage Family members are herbs and sometimes shrubs, with rough, hairy alternate leaves with smooth edges. The completely symmetrical flowers have united petals and appear in coiled clusters which unbend as the flowers become older. There are usually 5 sepals, 5 petals and 5 stamens. The 4-parted ovary produces 4 nutlets.

The Forget-me-nots of Victorian valentines and old-fashioned gardens belong to this family. We have 5 genera and 13 species in our range, all native.

Common Fiddleneck

Amsinckia intermedia F. & M.
Borage Family. BORAGINACEAE

Dede Gilman

Common Fiddleneck is a slender, bristly annual up to 30 inches high with linear leaves up to 6 inches long. The orange-yellow flowers are tightly coiled in a fiddleneck and have a narrow tube which expands into flat lobes. The style is thread-like with a 2-lobed stigma.

Common Fiddleneck is abundant on open, grassy hillsides and burns. It blooms from February to May.

Rough as the plants are to touch, cattle enjoy them, and in Arizona they are known as *Sacate Gordo,* Spanish for "fat grass."

Wilhelm Amsinck (1752-1831) was an early 19th century patron of a botanic garden in Hamburg. *Intermedia* is translated as "intermediate." There are many species of this plant and this one was probably seen as halfway between two others.

Popcorn Flower

Cryptantha spp.
Plagiobothrys spp.
Borage Family. BORAGINACEAE

Barry Silver

The small, white, Forget-me-not-like flowers are separated into genus and species by very technical inspections of the nutlets they produce as fruit. They are all bristly, hairy, little plants a foot or so high with narrow leaves, most of them basal. Eight species of these two genera have been found in the Santa Monica Mountains. They are very difficult to identify.

Popcorn Flower blooms from March until July and can be found throughout the range. Because they bloom so abundantly and look like a light snowfall when they cover a field, the Spanish speaking people once called them *nievitas*, "little snow."

The juice from the stems can stain your fingers, and the roots of some species contain a rich purple dye.

Plagiobothrys is made up of two Greek words meaning "hollow at the side," an allusion to the pitted face of the nutlets. *Cryptantha* means "hidden flower." The flower of the first described species had small inconspicuous flowers which self-fertilized without opening.

Wild Heliotrope, Chinese Pusley

Heliotropium curassavicum L. var. *oculatum* (Heller) I.M. Johnst. ex Tidestrom.
Borage Family. BORAGINACEAE

Geoffrey Burleigh

Wild Heliotrope is a somewhat fleshy perennial with spreading stems, 6 to 12 inches high. The alternate, succulent leaves are wedge-shaped, ¼ to 1½ inches long. The flowers are white, purplish near the center with yellow spots in the throat. Each one is a long, slender tube with 5 expanded lobes. There are 5 sepals and 5 stamens. The whole flower cluster forms a tightly coiled spike.

Wild Heliotrope is frequent near the coast and in damp places throughout. It blooms from April to October.

The Spanish Californians are said to have blown a dry powder made from the mucilaginous leaves into the wounds of men and animals.

Heliotropium is from Greek meaning "sun-turning," referring to the summer solstice when the first described species bloomed. *Curassavicum* is a place name coming from Curaçao, an island in the Dutch West Indies. One of the first collections was made there.

Mustard Family. BRASSICACEAE (CRUCIFERAE)

The Mustard Family is one of the easiest of all plant families to recognize because of the 4 distinct petals arranged to form a cross. This family is also known as CRUCIFERAE because of this feature. There are 4 sepals and 6 stamens, two of which are shorter than the other four. The stem leaves are alternate and many species have basal rosettes. The mature seed pods (capsules) take several distinct shapes, but are always noticeable. They are usually needed if uncertain about which species is being examined.

This very large family contains a number of our common food plants—cabbage, turnip, radish, etc. There are 27 genera locally, a great many of them introduced weeds. Watercress and Sweet Alyssum have crept in from elsewhere. Even the abundant Wild Radish *(Raphanus sativus)* frequent in waste places and the Sea Rocket *(Cakile maritima)* on the sandy beaches are immigrants.

Tower Mustard

Arabis glabra (L.) Bernh.
Mustard Family. BRASSICACEAE (CRUCIFERAE)

Tower Mustard is an annual or biennial with bluish-green stems 2 to 4 feet high. They are erect and seldom-branched, smooth and covered with a waxy, white substance that quickly rubs away. The broadly lance-shaped leaves are attached at the narrow end and most of them grow in a rosette at the base. The tiny, yellowish-white petals exceed the sepals, but just barely. The many pods are erect and straight, 3 to 4 inches long.

Tower Mustard blooms from March to May below 2000 feet on the seaward side of the mountains.

There are 21 species of *Arabis* in Southern California. The name for the whole genus is Rock Cress. The only other species reported from the Santa Monicas is the perennial, *A. sparsiflora* var. *californica* with purple flowers and leaves that do not grow in a basal rosette. It may be encountered in Griffith Park.

The Greek word *Arabis* was used interchangeably with *Cardamine, Iberis* and the Latin *Nasturtium,* which all mean "mustard" or "cress." *Glabra* means "smooth."

Black Mustard
Brassica nigra (L.) Koch.
Mustard Family. BRASSICACEAE (CRUCIFERAE)

James P. Kenney

Black Mustard is an erect, branching annual reaching 3 to 6 feet, and sometimes 12 feet tall. It is slightly hairy when young, becoming smooth with age. The lower leaves are parted with a large lobe at the end. The upper leaves are much smaller. The bright yellow flowers appear on small stems at the ends of the branches. The narrow pod, almost an inch long and ending in a round beak, presses against the main stem.

A ubiquitous grain weed from the Old World, Black Mustard colors entire fields an impressive bright yellow in the spring. It blooms from March to July.

Black Mustard is cultivated for its seeds which are the source of the condiment. It is thought to have been introduced here and encouraged by the Franciscan Padres, who are said to have scattered its seeds freely as they walked along the Camino Real to make the road easier to discern. Helen Hunt Jackson, the author of *Ramona,* described it as a "golden snowstorm."

Brassica rapa ssp. *sylvestris* is also plentiful in our mountains. The leaves of this species are broader and the pods spread away from the stems.

Brassica is the Latin name for "cabbage." The color of the seeds is responsible for the "black" *(nigra)* part of the name.

Sea Rocket

Cakile maritima Scop.
Mustard Family. BRASSICACEAE (CRUCIFERAE)

Linda Hardie-Scott

The Sea Rocket is a smooth-skinned, succulent annual with prostrate, spreading stems. The leaves are deeply parted into oblong lobes with rounded tips. They are 1½ to 3 inches long. The small petals, less than ½ inch across, are pink to purplish. The fruit is a swollen two-jointed pod.

This native of Europe and North Africa blooms from March to October on beach sands from Point Mugu to Point Dume.

Sea Rocket thrives, forever extending its range, as a result of its unique seed dispersal system. The upper part of the fruit has a single seed sealed inside a corky outer covering which breaks off when ripe. This capsule is impervious to salt water and floats off on the current to new locations. The lower half of the fruit also contains a seed which remains attached to the parent plant and produces a new Sea Rocket on its already proven favorable home ground.

Cakile is the old Arabic name for this plant. *Maritima* means "of the sea."

Shepherd's Purse
Capsella bursa-pastoris (L.) Medic.
Mustard Family. BRASSICACEAE (CRUCIFERAE)

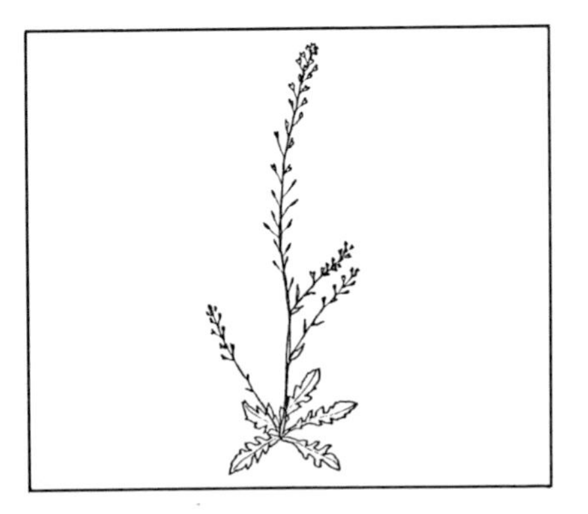

Shepherd's Purse is an erect annual 3 to 18 inches tall with simple or branching stems and basal leaves in a rosette, deeply cut with a large lobe at the end. The upper leaves are toothed and clasp the stem with ear-like lobes. The very small, white flowers are borne on short stems. The pods are quite distinct—heart-shaped, flattened and containing many seeds.

Shepherd's Purse, a cosmopolitan weed from the Mediterranean region, is common in waste places and disturbed areas away from the immediate coast. It blooms and sets its distinctive seeds the year round.

The fruits have a peppery taste and the whole plant has been used as a remedy for many ills. A common name in Europe is Mother's Heart and children are said to have played a game of pulling the seed pouch with a friend until it splits and then taunting, "You've broken your mother's heart!"

In Latin *Capsella* means "a little box" and *bursa-pastoris* means "a shepherd's purse."

Milkmaids, Toothwort
Cardamine californica (Nutt.) Greene.
Mustard Family. BRASSICACEAE (CRUCIFERAE)

Linda Hardie-Scott

Milkmaids is an erect perennial 6 inches to 2 feet high. The basal leaves are mostly roundish with 3 leaflets. The stem leaves are deeply lobed and may have from 3 to 5 pinnate leaflets. The white to pale rose flowers have the standard Mustard Family structure. The pod is slender, an inch to 1½ inches long.

Milkmaids bloom from February to April on moist, shady banks in Oak Woodland and Chaparral away from the ocean.

One of the common names, Toothwort, refers to the bulges or teeth on the roots. Wort is an old English name for plant or herb. The little underground tubers have a sharp taste and in other areas this early appearing wildflower is called Pepper Root.

Cardamine signifies "I strengthen the heart"—because of some supposed medicinal value.

Western Wallflower
Erysimum capitatum (Dougl.) Greene.
Mustard Family. BRASSICACEAE (CRUCIFERAE)

Geoffrey Burleigh

The Western Wallflower is an erect, unbranched plant 6 to 18 inches high. The roughish leaves are several times longer than wide and toothed when appearing at the base. The stem-leaves are smooth-edged and much smaller. The flowers grow on short stalks in a loose cluster at the ends of the main stem. They vary in color, but are usually deep orange. The four petals are narrow at the base and are ½ to ¾ inch long. The 4-sided pod is 2 to 4 inches long.

Western Wallflower blooms from March through May in openings in Chaparral below 2000 feet.

The name Wallflower does not seem appropriate until you know that in the Old World a closely related plant is common on dry stone walls. Our Wallflower is very fragrant.

Erysimum comes from Greek and means "to help or save." Some of the species were supposed to have had medicinal value. *Capitatum* refers to the way the flowers form a cluster.

Shining Peppergrass
Lepidium nitidum Nutt.
Mustard Family. BRASSICACEAE (CRUCIFERAE)

The Shining Peppergrass is an annual 1 to 16 inches tall with the stem branching near the base and mostly smooth herbage. The divided leaves are 1/2 to 4 inches long with an enlarged terminal lobe. The teeny, white petals narrow to a distinct claw. The shiny, circular pods, about 1/6 inch long, have a narrow margin notched at the top.

Shining Peppergrass grows on open slopes in Grassland, Coastal Sage and on rocky outcroppings throughout. It blooms from February to May.

Dense colonies of this plant in the pod stage can color their area reddish in late spring. The small, round, flat pods have a peppery taste.

Lepidium in Greek means "a little scale" from the shape of the pods. *Nitidum* is from Latin meaning "shining," also referring to the pods.

Wild Radish
Raphanus sativus L.
Mustard Family. BRASSICACEAE (CRUCIFERAE)

Arline Denny

The Wild Radish is an annual or biennial plant which first forms leaves at ground level, then the flowering stem which grows up to 4 feet tall. The lower leaves are deeply cut with a large terminal lobe. The upper ones are merely toothed. The numerous, showy flowers are on small stems at the end of the main one. They can be yellow, lavender, pink or white with purple veins on the petals. The cylindrical, pointed pods are spongy, 1½ to 3 inches long with 2 to 8 seeds.

Wild Radish is abundant in waste places everywhere and blooms from February to July.

The Wild Radish is the common garden variety naturalized from Europe and gone back to the wild. In some areas it grows in such masses that it is an influential factor in the color scheme of the landscape. The tap root which we eat becomes tough and inedible as the plant reaches its flowering stage.

Raphanus is Greek for "quick appearing" because of the rapid germination of the seeds. *Sativus* means "sowed," indicating the plant is a cultivated one.

Watercress

Rorippa nasturtium-aquaticum (L.) Schinz & Thell.
Mustard Family. BRASSICACEAE (CRUCIFERAE)

James P. Kenney

Watercress is a perennial herb with weak, mostly prostrate stems that root at the nodes. The leaves are pinnately parted into 3 to 9 leaflets ¼ to 4 inches long. They are light green and peppery in flavor. The white flowers are only ⅛ inch wide and appear on tiny stems at the ends of the branches. The pods are small and bean-like with 2 rows of seeds.

Watercress, a native of Eurasia which has naturalized itself all over North America, is found in quiet water or along streambanks throughout the area. It blooms from May to October.

This is the common Watercress used in salads. It is rich in ascorbic acid, vitamin C. If you collect it in the wild, it is well to be sure that the stream it is growing in or near is not polluted.

Rorippa is from an old Anglo-Saxon word whose meaning has been lost to time. The garden nasturtium belongs to a totally different family but the name is scientifically applied to a number of the cresses as well and means "twisted nose" in reference to the pungent qualities. *Aquaticum* means "found in the water."

London Rocket

Sisymbrium irio L.
Mustard Family. BRASSICACEAE (CRUCIFERAE)

London Rocket is a smooth, erect annual 1½ to 2 feet tall. The leaves are deeply pinnate with a large, terminal lobe. The pale yellow flowers only ⅛ inch long appear at the top. The elongate, linear pods, none more than 1¼ inches long grow up and out from the stems.

London Rocket grows best in disturbed places and blooms from March to October.

The common name of the whole genus is Hedge Mustard. Not one of the four members of this genus in our mountains is a native. They have come to us from Europe, North Africa and the near East. *S. officinale* has fruits only ½ inch long which press close to the stem. Tumbling Mustard *(S. altissimum)* and Oriental Hedge Mustard *(S. orientale)* have pods 2 to 4 inches long, but the first has very finely divided leaves while the second one has leaves that are almost entire.

Sisymbrium is the Greek name for some plant of the Mustard Family. *Irio* refers to "a kind of cress."

Golden Prince's Plume, Desert Plume

Stanleya pinnata (Pursh) Britton.
Mustard Family. BRASSICACEAE (CRUCIFERAE)

Linda Hardie-Scott

The Prince's Plume is a coarse, erect perennial 2 to 5 feet tall with more or less erect branches. The pale gray, smooth leaves, up to 6 inches long, may be either deeply divided or not at all. The bright yellow flowers appear in a showy terminal cluster. The fruit is a narrow curved pod on the end of a short stem.

This showy plant, frequently encountered in the deserts but rarely in our mountains, appears on road cuts facing the ocean near Malibu Lagoon and in Cheeseboro Canyon. It blooms all year.

It is likely that desert Indians used the young leaves for greens but rinsed them thoroughly since they carry some of the poisonous mineral selenium. It is a scarce and relatively recent arrival in the Santa Monicas so it is unlikely that natives here found it useful or even found it.

The genus is named for Lord Edward Smith Stanley, 13th Earl of Derby (1773-1849), ornithologist and once president of the Linnean Society (1828-1833). *Pinnata* means "feathered," in reference to the leaves.

Hairy Fringe-pod, Lace-pod

Thysanocarpus curvipes Hook.
Mustard Family. BRASSICACEAE (CRUCIFERAE)

James P. Kenney

The Hairy Fringe-pod is a slender annual 1 to 1½ feet high. The hairy, basal leaves are narrow at the base with somewhat triangular lobes. The upper leaves are less wide and clasp the stem with ear-like lobes. The white flowers are minute and inconspicuous. The fruits, which hang down, have a membranous border with radiating veins and overlapping hairs in the center.

Hairy Fringe-pod is fairly common in open rocky areas in Chaparral. It blooms from February to April, but the attractive fruits hang on longer.

Narrow-leaved Fringe-pod *(T. laciniatus)* is frequently encountered throughout the mountains below 2000 feet. The upper leaves are not lobed, nor are they, or the fruits, as noticeably hairy as *T. curvipes.*

Thysanocarpus is a composite of two Greek words meaning "fringed fruit." *Curvipes* means "curved feet."

Cactus Family. CACTACEAE

The Cactus Family plants are peculiar fleshy perennials more or less armed with bristles and spines, and rarely having leaves. The family is easily recognized but individuals may be difficult to place precisely. There are about 124 genera and 1,235 species, nearly all natives of America—many with very showy flowers.

Coastal Prickly Pear

Opuntia littoralis (Engelm.) Ckll. var. *littoralis.*
Cactus Family. CACTACEAE

Geoffrey Burleigh

The Coastal Prickly Pear is a sprawling plant with broad, flat joints usually forming thickets up to 5 feet high. The joints have spines coming from specialized points called areoles. The showy flowers are pale yellow with many overlapping, waxy petals 1½ to 2 inches long. There are numerous stamens. The fruits are pear-shaped, dark reddish-purple and covered with bristles.

Coastal Prickly Pear is found in Coastal Sage Scrub at low elevations along the coast. It blooms from May to June.

The fruits, called *tunas* by the Spanish Americans, are edible. Relished by the Indians, they are still eaten today. Great care and some skills are required to remove all of the bristles. After removing the spines and skin, the fleshy pads, called *nopales*, are sliced and eaten. If you wish, you can make Cactus Candy by soaking ½ inch slices of the *tunas* over-night in cold water and simmering slowly in a syrup of 3 cups sugar, ½ cup water, 2 tbs. orange juice, and 1 tbs. lemon juice until the syrup is almost absorbed.

O. oricola differs from *O. littoralis* by having yellow rather than white spines. The two can be found growing together in Malibu. *O. prolifera* has only recently been found growing at one or two scattered locations. A fourth species, *O. basilaris,* has been located at Wildwood Park, Thousand Oaks.

Opuntia is an old Latin name used by Pliny, formerly belonging to another plant. *Littoralis* means "of the seashore."

Caper Family. CAPPARACEAE

Members of the Caper Family are heavily scented herbs or shrubs with alternate leaves composed of 3 or more leaflets on stalks. The flowers have 4 sepals, 4 petals, 4 or more stamens of equal length and a superior ovary. The fruit is a 2-valved capsule often on a slender stalk. There are about 30 genera and 650 species in warm temperatures and tropical regions. The capers for our dinner table come from plants in this family.

Bladderpod

Isomeris arborea Nutt. var. *arborea.*
Caper Family. CAPPARACEAE

Dede Gilman

The Bladderpod is a shrub from 3 to 10 feet high with hard, yellow wood and a characteristic disagreeable smell in leaf and blossom. The leaves are narrowly thrice-divided. The showy flowers are dull yellow with protruding stamens and are borne on small stalks at the ends of the branches. The large leathery much-inflated seed vessels, like fat pea pods, droop on long stalks and are a prominent feature.

This is a Southern California plant which is at home on seashore bluffs as well as dry desert sands, and may be found in bloom throughout the year. It is the only species of the genus.

Another common name is Burro Fat given because of its strong odor.

The generic name is from the Greek *isos,* "equal" and *meris,* "part," describing the equally divided pod. The species name is derived from the Latin from "tree" referring to the tree-like habit of the shrub.

Honeysuckle Family. CAPRIFOLIACEAE

The plants of the Honeysuckle Family are either shrubs or vines with usually simple, opposite leaves. The flowers are in 5 parts with the petals joined making a tubular corolla which often flares at the end. There are 4 to 5 distinct stamens and an inferior ovary. The fruit is a fleshy berry.

There are about 14 genera and about 400 species in the Northern Temperate Zones. Many are grown as ornamentals.

Southern Honeysuckle

Lonicera subspicata H. & A. var. *johnstonii* Keck.
Honeysuckle Family. CAPRIFOLIACEAE

Tim Thomas

Southern Honeysuckle is a straggling, evergreen shrub, 3 to 8 feet high with thin, shredded, grayish bark on the old branches. The leaves are rounded, about an inch long and ½ inch wide. The cream-colored flowers are in short, terminal spikes. The flower tubes are about ½ inch long, 2-lipped and 5-lobed with the lobes curving back. There are 5 sepals and 5 stamens. The fruit is a yellowish to reddish berry.

Southern Honeysuckle blooms in April and May in Chaparral throughout the area.

The flowers are often quite fragrant. Southern Honeysuckle is a rapid grower, propagates easily from cuttings and prefers the north side of the house. *L. hispidula,* the other species in our mountains, has reddish-purple flowers and occurs in one remote location.

This genus was named for Adam Lonitzer (1528-1586), a German herbalist, physician and botanist. He wrote a standard herbal, titled *Kreuter-buch,* that was in print for almost 250 years. He was also a proofreader for his father-in-law who specialized in printing revisions of old herbals. *Subspicata* means "somewhat spiked."

Mexican Elderberry

Sambucus mexicana Presl.
Honeysuckle Family. CAPRIFOLIACEAE

Dede Gilman

Mexican Elderberry is a large shrub or small tree up to 20 feet high, sometimes even taller. The leaves are opposite, pinnately compound with ovate, finely-toothed leaflets. The small white flowers are in flat-topped compound clusters. The fruit is a blue berry.

Mexican Elderberry can be found in Chaparral and open washes throughout the mountains. It blooms from April to August.

Geoffrey Burleigh

Indians found this a most versatile plant. They not only ate the fruit, they made a soothing tea from the dried blossoms for fevers or spread it externally on sprains and itches. After pushing out the pith in the wood, they turned it into flutes or clappers. They also used the wood to make bows. The fruit can be used for pies and jams.

Sambucus is from a Greek word for a musical instrument made from alder wood. From *mexicana* we learn that it is also native to that country.

Creeping Snowberry
Symphoricarpos mollis Nutt. in T. & G.
Honeysuckle Family. CAPRIFOLIACEAE

James P. Kenney

Creeping Snowberry is a low straggling shrub less than 18 inches high with stems several feet long. The downy, roundish leaves are ½ to ¾ inch long with various margins. The pinkish flowers are about ⅛ inch long and appear in clusters. The fruit is a berry with two nutlets.

Creeping Snowberry is found in Chaparral, Oak Woodland and along streambeds throughout. It blooms in April and May.

The soft, pure-white fruits are very evident in fall and winter after the deciduous leaves have been shed. *S. mollis* can be propagated from divisions and serves as a good north side ground cover in the garden.

The generic name is composed of two Greek words meaning, "fruit borne together"—because of the clustered berries. *Mollis* is from Latin meaning, "soft, tender, flexible."

Pink Family. CARYOPHYLLACEAE

The plants of the Pink Family are herbs with swollen stem joints and simple, opposite leaves. The petals and sepals are usually 5 each. The stamens are either the number of the sepals or twice as many. The ovary is superior and the fruit is a capsule.

Many of the members of this family including the familiar Carnation are cultivated in gardens. There are nine genera in the Santa Monica Mountains with ten native and six introduced species.

Windmill Pink
Silene gallica L.
Pink Family. CARYOPHYLLACEAE

Dede Gilman

Windmill Pink is an annual herb 10 to 15 inches high. The opposite leaves are spade-shaped up to 1½ inches long with a short, abrupt tip. All of the herbage is hairy and sticky. The 5-toothed urn-shaped sepals are constricted at the top. The 5 white to pink oval petals are clawed and slightly twisted. There are 10 stamens. The fruit is a capsule.

Windmill Pink is frequent in weedy and disturbed ground throughout and blooms from February to May.

S. gallica is a native of Europe but is now so widely distributed from near the sea to inland that it is hard to believe that it has not always been here. The common name is derived from the fact that each petal is turned at a slight angle making the flower look like a tiny windmill.

Gallica means "French."

Fringed Indian Pink

Silene laciniata Cav. ssp. *major* Hitchc. & Maguire.
Pink Family. CARYOPHYLLACEAE

Sarah Thomas Schwaegler

Fringed Indian Pink is a sticky, hairy perennial with many stems, 6 to 16 inches high, growing from a taproot. Each of the 5 scarlet petals is deeply slashed into 4 narrow lobes. They are ½ to ¾ inch across. The sepals are urn-shaped. There are 10 stamens and a 3-parted pistil. The leaves are opposite and tend to be linear. The fruit is a capsule.

Fringed Indian Pink is common in Coastal Sage and Chaparral throughout and blooms from April until July.

One popular name is Campion, meaning field flower.

The stickiness of the stems and leaves can trap small insects so this and similar plants are often called Catchfly. It has been suggested that all of this gummy hairiness as well as the stickiness on the outside of the tough, tubular flowers is designed to discourage ants which are not very desirable visitors for promoting cross-fertilization. Ants' progress is slow, and they cannot visit many plants far apart compared to the flying insects which are attracted to the bright color and smooth insides of the flower and which are well-adapted for bringing pollen from distant plants.

Another species, *S. californica,* is very similar but does not occur this far south.

Silenus, the foster father of Bacchus, was supposed to be covered with foam. Many plants of this genus do have a viscid secretion. *Laciniata* means "torn" and refers to the fringed petals.

Spurrey

Spergula arvensis L.
Pink Family. CARYOPHYLLACEAE

Linda Hardie-Scott

Spurrey is an annual 4 to 16 inches high. The thread-like leaves, 1¼ inches long, are whorled. The small, white flowers are in an open cluster and open only in sunlight. There are 5 sepals, 5 petals, 5 or 10 stamens and 5 styles. The fruit is a 5-valved capsule.

Spurrey is locally abundant on sandy flats near the coast and blooms in March and April.

This introduction from Europe has become widespread throughout the eastern states, Canada and the Pacific Northwest. In California, it is coastal and shows up in fields, orchards and gardens. It has a number of interesting common names: Devil's Gut, Sandweed, Pick-purse, and Yarr.

Spergula is from Latin, *spargere,* "to sow seeds" to produce forage. *Arvensis* means "of the fields."

Common Chickweed
Stellaria media (L.) Vill.
Pink Family. CARYOPHYLLACEAE

Common Chickweed is an annual blooming through the winter in sheltered places. The stems are slender, weak and trailing. The opposite, oval leaves have smooth edges and are ¼ to 1 inch long. The ¼ inch white flowers are single in the axils of the leaves. The 4 to 5 petals are 2-parted at their tips and surrounded by 5 slightly long sepals. The fruit is a many-seeded, dry, oval capsule.

Common Chickweed is a weed of waste places and a pest in cultivated areas, blooming from February to April.

Common Chickweed is native of Eurasia. Mouse-ear Chickweed *(Cerastium glomeratum),* also a Eurasian immigrant, is much less common. Hairy stems and leaves and an elongate capsule differentiate it from the Common Chickweed.

Stellaria is from Latin, *stella,* "a star" because of the shape of the flowers. *Media* means "the middle." This species is midway in size between two others that are not native here.

Goosefoot Family. CHENOPODIACEAE

The plants of the Goosefoot Family are herbs or shrubs, often succulent or scurfy, with simple alternate leaves. The flowers are small, greenish and lack petals. Members of this family are widely distributed, preferring for the most part saline soils near the ocean or on deserts. Beets and spinach belong to this family, but so do the troublesome Pigweeds, Russian Thistle and Lamb's Quarters.

The fruit is most often small, thin-walled and one-seeded.

Of the more than 100 genera known world-wide, we have eight in the Santa Monicas.

Coast Saltbush

Atriplex lentiformis (Torr.) Wats. ssp. *breweri* (Wats.) Hall & Clem.
Goosefoot Family. CHENOPODIACEAE

Coast Saltbush is a wide, much-branched shrub 3 to 10 feet high. The oval leaves are entire, 1 to 2 inches long, broad at the base and sometimes pointed at the tip. The small, greenish flowers of both sexes cluster along the branches. The fruits form flattened disks ¼ inch long.

A. lentiformis is frequent in Coastal Sage. It blooms from July to November.

Over 30 species of *Atriplex* occur in the state, 8 of them reported in the Santa Monicas. The two others you'll see most often are Australian Saltbush *(A. semibaccata)*, which is low, spreads like a groundcover and displays small bright red fruits, and the annual, *A. patula* ssp. *hastata,* which has distinctive leaves shaped like an arrowhead with the basal lobes pointing outward. All of the *Atriplex* feel mealy when you rub the leaves.

Another common name for this plant is Lenscale.

Atriplex is the ancient Latin name for these plants. *Lentiformis* means "shaped like a lens," referring to the fruits.

Five Hook Bassia

Bassia hyssopifolia (Pall.) Kuntze.
Goosefoot Family. CHENOPODIACEAE

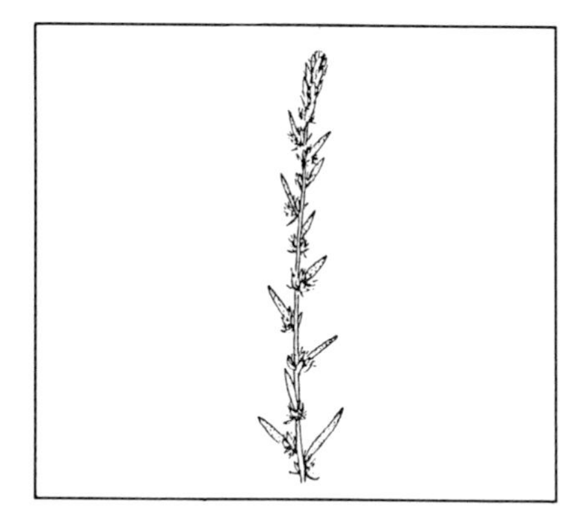

Five Hook Bassia is a gray, soft, hairy annual with stems branching from the base. The leaves are linear, alternate and ¾ to 1½ inches long. The flowers are very small and are crowded in the leaf axils. The small seed is enclosed in a 5-hooked hairy calyx. Dense clusters of the 5-hooked hairy calyx give the slender branches a woolly appearance.

A Eurasian native, the Five Hook Bassia is common in disturbed ground near the coast, blooming from May to September. This plant likes alkaline soils and is most abundant in salt marshes.

Ferdinando Bassi (1710-1774) was an Italian botanist and the Prefect of the Bologna Botanic Gardens. *Hyssopifolia* means "leaves like *hyssop,*" an aromatic herb in Greece. *Folia* means "leaves."

Mexican Tea
Chenopodium ambrosioides L.
Goosefoot Family. CHENOPODIACEAE

Linda Hardie-Scott

Mexican Tea is an erect herb 2 to 3½ feet tall with rigid, hairy branches. The 2 to 5 inch long leaves are often sticky, hairy and wavy on the edges. The tiny, greenish flowers cluster up and down the branches. They have 3 to 5 sepals which enclose the dry fruit.

This tropical American native is common in moist places throughout and blooms from July to February.

The genus includes plants known as Pigweed, Lamb's Quarters, and Goosefoot. Rub the leaves for the mealy feel, like that of the *Atriplex*. *C. album, C. multifidum,* and *C. murale* are the most common of the 10 other species inhabiting the Santa Monicas.

Chenopodium is from two Greek words meaning "goose" and "foot" which refers to the shape of the leaves in some species. *Ambrosioides* means "like the genus *Ambrosia*."

Pickleweed, Glasswort
Salicornia virginica L.
Goosefoot Family. CHENOPODIACEAE

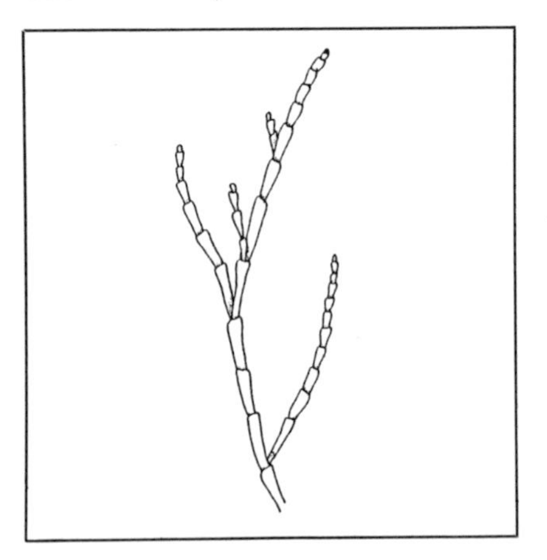

Pickleweed is a succulent, leafless, perennial with jointed stems and opposite branches. The minute flowers are sunk into the upper joints. When trying to detect the flowers, look for a few small stamens just sticking up from the fleshy spike. The seed is 1/16 inch long.

S. virginica is locally abundant in salt marshes and the upper parts of beaches along the entire coast. It blooms in late summer from August to November. In the fall, Pickleweed turns red and bronze and even purple, or olive.

Cattle find *S. virginica* tasty. *S. europaea,* another local species, is a bushy annual and much less common. It is made into pickles in Great Britain.

Salicornia is from two Greek words for "salt" and "horn." They are saline plants with hornlike branches. *Virginica* tells us that the first specimen of this species was collected from that state.

Russian Thistle
Salsola iberica Sennen & Pau.
Goosefoot Family. CHENOPODIACEAE

Dede Gilman

Russian Thistle is a bushy annual plant that usually breaks off at the ground line when mature to form a "tumbleweed." The slender, reddish stems, up to 10 inches high, are tender when young but become stiff and spiny in age. Young leaves are an inch long, narrow and pointed. They drop off and the next ones are stiffer, triangular and end in spines. The flowers appearing from base to tip are without petals. They are small, greenish, 5-parted with a veiny wing which nearly covers the fruit. The fruit is dry in maturity and has a broad papery border.

This native of the plains of southeastern Russia and western Siberia is abundant throughout disturbed places and blooms from May to November.

Fifty years ago Russian observers reported stretches of this plant for 500 to 600 miles in the most fertile parts of their country and said that it had driven every farmer out of the area. It came to the United States in 1873 in imported flax seed sown in South Dakota. It is a true "tumble weed" and has tumbled to California prolifically. A single plant may produce from 20,000 to 50,000 seeds and scatters them as it tumbles.

Salsola is from Latin, *salsus,* "salty." *Iberica* indicates it might first have been described from a species collected in Spain.

California Sea Blite
Suaeda californica Wats.
Goosefoot Family. CHENOPODIACEAE

California Sea Blite is a spreading, perennial subshrub 2 to 7 feet across and woody only at the base. The fleshy, linear leaves ½ to 1¼ inches long are densely crowded on the stems. The small greenish flowers are clustered on terminal branches. The 5 stamens are exserted. The fruit is thin-walled and one-seeded.

This species of *Suaeda* is found occasionally in salt marshes and along beaches from one end of the coast to the other. It blooms from August to October.

The only other species of this genus in our mountains, *S. depressa*, is a rare annual found near the ocean at Pacific Palisades.

The common name "blite" has been applied to many members of the Goosefoot Family for centuries. It is not at all kin to the English word "blight." Instead it means "insipid" in Latin and indicates that few of these plants are tasty.

The Indians used the seeds along with many others for a mush called *pinole*. Those who have duplicated this dish assert that it is indeed insipid. The plant made a rich, black dye for the weaving of Indian baskets.

Suaeda is an Arabic name of antiquity.

Rock-rose Family. CISTACEAE

The plants of the Rock-rose Family are shrubs or herbs with leaves that occur singly and without divisions. The flowers have 5 petals that drop early, 2 small outer bractlike sepals and 3 inner, larger ones that persist when all others are gone. There are many stamens. The ovary is superior with a one-parted style. The fruit is a capsule with tiny seeds.

We have only one native and one introduced species in the Santa Monicas.

Rock-rose
Cistus villosus L.
Rock-Rose Family. CISTACEAE

Linda Hardie-Scott

The Rock-rose is an erect shrub about three feet high. The opposite, oval leaves are 1½ to 2½ inches long. The deep pink flowers have 5 petals each about an inch long. There are 5 oval sepals with pointed tips. The fruit is a capsule with shiny brown seeds.

Rock-rose has often been planted in semi-wild areas as it is known to grow vigorously in hot sun and poor soil and will even flourish in ocean spray. It is particularly abundant at Will Rogers State Park, blooming in June and July.

This is but one member of a number of species of *Cistus* which have become naturalized. You may encounter a very attractive pure white one, *C. corbariensis.*

The myrrh brought by the Magi to Bethlehem was a fragrant gum resin from *C. villosus*. It may still be obtained today in specialty shops to add long-lasting scent to dried flower mixtures.

Cistus is an ancient Greek name. *Villosus* means "hairy."

Common Rock-rose
Helianthemum scoparium Nutt. var. *vulgare* Jeps.
Rock-Rose Family. CISTACEAE

Barry Silver

The Common Rock-rose is a low, tufted perennial up to a foot high with many ascending, spreading stems and narrow, linear leaves. The yellow flowers appear in a compound cluster and drop early. Each of the 5 petals is about ¼ inch long. The 2 outer sepals are linear and the 3 inner ones are rounded. There are numerous stamens.

Common Rock-rose blooms from March to October on dry, sunny slopes usually on the coastal side of the range throughout.

This Rock-rose (no relation to the Rose Family) is particularly abundant after fires. The broad petals are crumpled in the bud, not pleated or folded and they never get properly smoothed out afterward which gives them the look of crepe paper.

There is a pink-flowered species in Europe thought to be the Rose of Sharon referred to by King Solomon in the Bible.

Helianthemum translates from the Greek to "sunflower," given because the flowers open only in the sun. *Scoparium* means "broom-like" and refers to the plant structure.

Morning Glory Family. CONVOLVULACEAE

The plants of the Morning Glory Family are twining or trailing herbs with alternate leaves, although parasitic ones are leafless. The flowers are usually large and showy. The 5 petals are fused into lobed bells or trumpets. There are 5 overlapping sepals and 5 stamens. The ovary is superior and the fruit is a capsule with 4 or fewer large seeds.

Garden Morning Glories and the common Sweet Potato belong in this family. We have four native and two introduced genera in the Santa Monicas totalling 11 species. Bindweed *(Convolvulus arvensis)*, introduced from Eurasia, closely resembles our native *Calystegia.* Bindweed has bracts halfway down the stem while bracts of the native are close under the sepals. Dodder, looking like strings of orange threads draped over some of our shrubs, is a species of *Cuscuta,* a parasitic plant.

Wild Morning Glory

Calystegia macrostegia (Greene) Brummitt.
Morning Glory Family. CONVOLVULACEAE

Dede Gilman

Wild Morning Glory is a scrambling vine with arrow-shaped leaves up to 2 inches long. The white, trumpet-shaped flowers are 2 inches in length on flower stems 1 to 4 inches long. They exhibit noticeable purple stripes and become purple in age. The species is quite variable with many subspecies in Southern California.

This *Calystegia* blooms from February to May on brushy slopes throughout and is particularly frequent in Coastal Sage.

C. soldanella is very similar but is found on beaches. The introduced Bindweed *(Convolvulus arvensis)* is also a look-alike but blooms from May to October when *C. macrostegia* is finished.

A tincture of the whole plant was once considered beneficial for medicinal purposes.

The generic name comes from two Greek words meaning "a covering cup." The species name means "a large covering." We seem to have here what would in literal translation give us a "large covering, covering cup."

California Dodder
Cuscuta californica H. & A.
Morning Glory Family. CONVOLVULACEAE

Barry Silver

The leafless Dodder plants have slender, thread-like, waxy, orange stems which coil about the host plant. They often form dense, colorful, tangled masses. There are flowers, but they are very tiny, with 5 white sepals and a 5-parted corolla. The absence of any green shows a true parasite. This species is parasitic on Sage, Buckwheat, Deerweed and Haploppapus.

Dodder is colorfully abundant in Coastal Sage and is scattered throughout at low elevations. It is most colorful in late spring and summer but occurs the year round.

The several species of *Cuscuta* are best identified by their host plants. The other 3 in the Santa Monicas can be found on Ceanothus, Cocklebur and Pickleweed. The several common names are interesting. It is known as Love-vine, Strangle-weed, Devil's Hair, Witch Hair and Golden Thread.

The name *Cuscuta* is of Arabic derivation meaning "dodder."

Stonecrop Family. CRASSULACEAE

Plants of the Stonecrop Family are succulent, fleshy herbs or shrubs. The flowers (rarely solitary) have sepals, petals and pistils which are all distinct and numbering the same whether from 3 to 12. The stamens are as many, or twice as many, as the petals. The fruit is a special kind of pod that opens only on one side.

Many species of this family are used in gardens because of the neat habit of the ornamental foliage. Hen and Chickens *(Echeveria* spp.) and *Kalanchoe* are favorite patio plants. Locally there are two genera, *Crassula* with one species and *Dudleya* with eight species.

Lance-leaved Dudleya
Dudleya lanceolata (Nutt.) Britt. & Rose.
Stonecrop Family. CRASSULACEAE

Linda Hardie-Scott

Lance-leaved Dudleya is a green, fleshy perennial. The plump, pointed leaves are in a basal rosette. The nearly naked flowering stalks may be up to 20 inches high with orange-red, waxy petals. The flowers are in coils near the end.

D. lanceolata is common on rocky slopes and banks throughout the mountains. It blooms from May to July.

There are seven other species of this genus in our mountains. They are difficult to identify and many of them are quite rare. Both this species and *D. pulverulenta* are excellent part-shade or full-sun, garden plants. They are best planted at a slight angle to avoid rot. These plants may be purchased from native plant nurseries. Collecting plants in the wild should not be done.

William Russel Dudley (1849-1911), a professor of botany, was the first head of the Botany Department at Stanford University (1892-1911), and was forestry editor of the Sierra Club magazine. *Lanceolata* means "lance-like," referring to the shape of the leaves.

Chalk Live-forever
Dudleya pulverulenta (Nutt.) Britt. & Rose.
Stonecrop Family. CRASSULACEAE

Geoffrey Burleigh

Chalk Live-forever is a vigorous plant densely covered with a white, chalky powder. The thick, large leaves, the outer ones up to 4 inches broad at the tips, all cluster near the ground in a symmetrical rosette sometimes 1½ feet across. There are small, scale-like leaves on the flowering stalks which are up to 16 inches high. The little, pale red flowers cluster on either side of this stalk at the top and never open wide. This beautiful plant can send up as many as eight flowering stems. The fruit is a pod, a special kind that opens only on one side.

Chalk Live-forever blooms from May to July and is common on rocky south or west facing slopes below 2000 feet throughout.

Pulverulenta means "powdery, dust-covered."

Gourd Family. CUCURBITACEAE

The plants of the Gourd Family are rapid-growing herbs, mostly vines with succulent stems and tendrils. The leaves are alternate, lobed and palmately veined. The flowers have 5 united petals, although there may be 6 or 7 in *Marah* (the one shown here), and are of 2 sexes on the same or different plants. The stamens are seemingly 3, but really 5 with two pairs united. The ovary is inferior and the fruit is a fleshy, berry-like structure with a rind and a spongy interior that is called a "pepo."

Melons, pumpkins and squash belong to this family as well as the gourds for which it is named. We have two genera in our mountains, *Marah* and *Cucurbita,* each with one species.

Calabazilla, Stinking Gourd

Cucurbita foetidissima HBK.
Gourd Family. CUCURBITACEAE

Dede Gilman

This is a rough, perennial vine with a large, carrot-shaped root. The stems with tendrils may stretch 15 feet along the ground and often root at their joints. The triangular leaves, 4 to 12 inches long are hairy below. The large, yellow flowers, both male with 3 stamens, and female with a 2 or 3-parted stigma, appear on the same plant. The fruit is a dry, round, smooth gourd 3 or 4 inches in diameter containing many seeds.

Calabazilla is a common weed in sandy soil blooming from June to September.

When driving in fairly flat areas you may see by the road something that resembles a group of abandoned and faded oranges where no citrus trees grow. Closer examination will show them to be the fruit of the Calabazilla. The foliage has a foul odor even when undisturbed although the flowers are reported to smell like violets.

Early Spanish settlers and Indians used the large root as a purgative or pounded it into a soap said to clean like nothing else could but needing thorough rinsing since any particles left in garments would irritate the skin. The Indians even ate the seeds, and Spanish ladies used the gourds as darning balls.

Cucurbita is the Latin name for "a gourd." *Foetidissima* means "very evil-smelling."

Chilicothe, Wild Cucumber
Marah macrocarpus (Greene) Greene.
Gourd Family. CUCURBITACEAE

Dede Gilman

Wild Cucumber is a trailing vine from a huge, fleshy root with long, stalked leaves palmately lobed and up to 4 inches across. The creamy flowers are interesting as some are male and some are female on the same plant. The male flowers appear on special stems in groups of 5 to 20 with the stamens exposed and noticeable. At the base of this special stem may be one female flower with a fat, little ovary which turns into a large, egg-shaped, bright green fruit at least 4 inches long covered with big soft green prickles that turn hard and spiny as the fruit dries. Inside are several handsome black seeds.

Chilicothe is one of the very first flowers to be noticed blooming from January to June. The 3-25′ stems sprawl vigorously over shrubs and ground everywhere, leaving a tangle of dried stems and leaves by midsummer. Because of its large root, it is one of the first plants to reappear after a fire. The long green stems spreading over the blackened soil are particularly noticeable.

The California Indians made necklaces of these seeds, polishing them by rubbing them along their oiled bodies. It is said that Indian children used them as marbles.

There is nothing at all edible about this plant even though it is called cucumber.

Another common name, Manroot, came about because the huge underground root is often large enough to resemble a human corpse.

The enormous root is intensely bitter and *Marah* refers to the bitter waters of that name in the Bible (Exodus 15:23).

> "And when they came to Marah, they could not drink of the waters of Marah, for they were bitter: therefore the name of it was called Marah."

Macrocarpus tells us that "the fruit is large" and it is.

Heath Family. ERICACEAE

Plants of the Heath Family are mainly evergreen shrubs and trees with thick, tough leaves that are alternate and simple. The flowers are united and shaped like an urn with 4 or 5 lobes. The stamens are either equal to, or double, the number of the lobes. The fruit is a capsule, berry or drupe.

Azaleas, Rhododendrons and Heathers belong to this family. We have two genera in the Santa Monicas—*Arctostaphylos* and *Comarostaphylis.*

Eastwood Manzanita

Arctostaphylos glandulosa Eastw. ssp. *glandulosa.*
Heath Family. ERICACEAE

James P. Kenney

This is a spreading shrub 2 to 4 feet high with several smooth, reddish, crooked stems from a basal burl which crown sprouts after fire. The oval or lance-shaped leaves, over an inch long and often an inch wide, are light to dark green, softly hairy on both surfaces and sharply pointed. The white flowers, ½ inch long, are urn-shaped. The red-brown fruit is round and sticky.

A. glandulosa may be found in open Chaparral away from the coast throughout flowering from January to March. It has been known to hybridize with our other species, Bigberry Manzanita *(A. glauca),* which does not have the basal burl and does not crown sprout after fire.

There are 50 species of Manzanita in North and Central America, 38 of them native to California.

To make an early western drink, scald a quart of ripe berries with a cup of boiling water, mash to a pulp, add a quart of cold water, allow to settle for an hour and strain. Sweeten to taste.

The common name is Spanish for "little apple" which the small fruits resemble. The generic name is from two Greek words, *arktos,* "bear," and *staphule,* "a grape," in reference to bears feeding on the grape-like fruits. It is also a favorite food of coyotes. *Glandulosa* means "provided with glands," secreting structures on the surface ending in hairs.

Spurge Family. EUPHORBIACEAE

Plants of the Spurge Family are herbs, shrubs or trees, usually with milky juice which is often poisonous. Many are succulent and cactus-like. The leaves are simple. The flowers have stamens or pistils, not both, and may or may not have petals and sepals. Showy bracts, as in cultivated Poinsettias and Crown-of-thorns, often simulate petals. Rubber comes from a plant in this family as does castor oil and tung oil.

There are over 280 genera of the EUPHORBIACEAE in the world but only 11 are native to California. We have three of these natives in our mountains and a fourth species is well naturalized.

California Croton

Croton californicus Muell.-Arg. var. *californicus*
Spurge Family. EUPHORBIACEAE

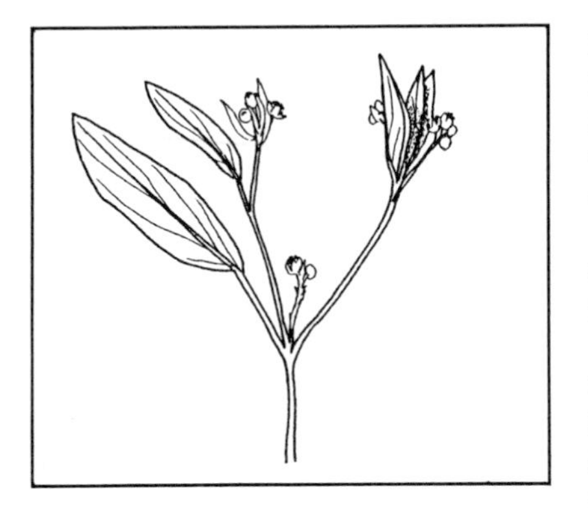

California Croton is a low perennial, slightly woody at the base, 2½ feet or less tall with gray, scurfy herbage. The oblong leaves are alternate, up to 2 inches long and have smooth margins. Flowers on one plant have pistils and those on another plant have stamens. The 5-parted petals and sepals are very tiny. The fruit is a capsule that explodes when dry.

Croton grows on sandy flats and slopes along the entire coast, flowering from April to November.

Unlike the California Croton, some other members of this family that can be found in our mountains have milky juice and the male and female flowes are borne on the same plant.

Croton is from a Greek word meaning "a tick." It is an old name for some members of this family because of the way the seeds look.

Turkey Mullein, Dove Weed

Eremocarpus setigerus (Hook.) Benth.
Spurge Family. EUPHORBIACEAE

James P. Kenney

Turkey Mullein is a low annual forming prostrate mats 1 or 2 feet across, from 1 to 2 inches to a foot tall. Stems and leaves are covered with forked, bristly hairs making the plant harsh to touch and light gray in color. The thick, oval leaves, ⅓ to 1½ inches long, are 3-nerved, on leaf stalks about the same length. The flowers are tiny. Those with 6 or 7 stamens have 5 or 6 sepals, no petals and appear in flat-topped clusters at the ends of the stem branches. The flowers with pistils consist only of a single hairy ovary and style. The fruit is a capsule with one, smooth, mottled seed.

Dove Weed is frequent on open grassy slopes and disturbed places throughout, blooming from June to November.

The common names reveal that the seeds are eaten by doves and turkeys. The stems and leaves contain a narcotic poison, and Indians threw pieces of these into streams to stun fish. The early Spanish did the same apparently as they called the plant *Yerba del Pescado,* "herb of the fish." The leaves resemble those of the true mulleins, *Verbascum* spp.

Eremocarpus is from two Greek words meaning "solitary fruit." *Setigerus* means "bearing bristles" referring to the hairy stems, sepals, ovaries and styles.

Rattlesnake Weed
Euphorbia albomarginata T. & G.
Spurge Family. EUPHORBIACEAE

Dede Gilman

Rattlesnake Weed is a many-stemmed, little, prostrate perennial that forms spherical mats on dry ground. The small, round leaves are heart-shaped at the base with a thin white edge. The lobes with white margins resembling petals form a cup. There are no true petals. The inconspicuous male flowers are clustered in the cup with a solitary female flower. The 12-30 male flowers consist of single stamens. The female flower has an elevated ovary pendant on a long stalk. The fruit is a capsule.

E. albomarginata is considered a weed and is common on dry slopes and in fields blooming from July to November.

Seven other members of this genus are recorded in the Santa Monicas. Two of them are introduced from other areas. Each of them can be identified as *Euphorbia* from the above description, but only *E. albomarginata* has the prominent white margins.

The early Spanish apparently called this and other species *Golondrina* which means "swallow." They all are reputed to be an antidote for rattlesnake bite when the leaves are pounded and bound wet on the wound.

Euphorbia commemorates Euphorbus, the physician of Juba II, the King of Mauretania. Juba II was educated in Rome and married the daughter of Anthony and Cleopatra. He wrote about one of the African cactus-like plants that was used as a powerful laxative and when he heard in 12 B.C. that Augustus Caesar had dedicated a statue to Antoninus Musa, Caesar's physician and the brother of Euphorbus, Juba decided also to honor his **own** physician, Euphorbus, by naming the plant that he had written about after him. The statue for Antonius has disappeared, but the name of the plant has come down through the centuries.

Castor Bean
Ricinus communis L.
Spurge Family. EUPHORBIACEAE

Linda Hardie-Scott

Castor Bean is a large herb or shrub 4 to 8 feet high with smooth, round, red stems. The large leaves are palmately lobed. The small greenish flowers are of two sorts on the same plant. The male flowers have numerous stamens and a 3 to 5-parted calyx. The female flowers have 3 red styles united at the base and a calyx that falls early. The fruit is a large, round, spiny capsule with shiny, mottled seeds.

It is the only member of this family with lobed leaves that grows wild in our mountains.

Castor Bean is common in disturbed ground or along roads throughout, blooming the year round. Since it is a native of warm regions of the Old World, it dies in low temperatures, so it has not become an important weed.

Although the seed of *R. communis* is the source of castor oil, long used as a laxative, it is toxic to humans and poisons livestock and poultry when eaten.

A Mediterranean sheep tick is named *Ricinus* and like most of this family, the seed resembles a tick. In Latin *communis* means "common, general."

Pea Family. FABACEAE (LEGUMINOSAE)

Members of the Pea Family are herbs, shrubs and trees with alternate, usually compound leaves. The flowers have 5 petals. The top petal is called the banner and is conspicuously larger than the other four. The two lower petals are joined into a keel and the two side petals form wings. This arrangement is called "papilonaceous" from Latin for "butterfly" and the garden Sweet Pea is a perfect example. The stamens and the single pistil are enclosed in the keel. The ovary is superior and the fruit, called a legume, is a pod which splits on both sides. This arrangement describes the subfamily, Papilonoideae, but not the other two subfamilies—the Mimosoideae, which includes Mimosa, Acacia and Mesquite, or the Caesalpinioideae, which includes Redbud, Palo Verde and Royal Poinciana.

This most significant plant family is distributed throughout the world with over 10,000 species—many of which are of great economic importance, providing food, forage, dyes and wood. Clover, alfalfa, senna, balsam, peanuts, indigo and all the peas and beans belong here.

We have 13 genera and 57 species in the Santa Monicas. Eleven of these species are clovers, 11 are in the genus *Lotus* and 13 are lupines.

California False Indigo

Amorpha californica Nutt.
Pea Family. FABACEAE (LEGUMINOSAE)

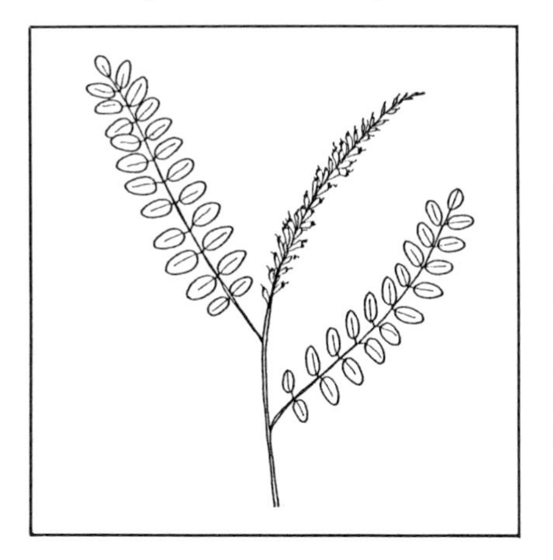

California False Indigo is a deciduous shrub 3 to 8 feet high. The leaves are 4 to 8 inches long, divided into 11 to 27 leaflets which are up to an inch long. The purplish flowers, which have only one petal, are crowded in upright spikes. The fruit is a small brown pod about ¼ inch long with 1 or 2 seeds.

This plant may be found below 2000 feet in Oak Woodland on the north side of the western portion of the range. It blooms from May to July. Locating this large shrub in bloom is exciting since it is unusual to find a flower with only one petal. The foliage has been described as sickeningly fragrant.

The generic name comes from a Greek word signifying "deformed"—referring to that one petal.

Southern California Locoweed

Astragalus trichopodus (Nutt.) Gray. ssp. *leucopsis* (T. & G.) Thorne.
Pea Family. FABACEAE (LEGUMINOSAE)

Steven R. Kutcher

Southern California Locoweed is a bushy, upright perennial with hairy stems a foot or so high. The leaves are pinnately divided into 10 to 15 pairs of oval leaflets each ½ inch or so long. The greenish-white flowers are in a 12 to 36-flowered cluster. Each flower appears on a short stem on either side of the main stem. The pods are thick, papery, inflated and covered with minute hairs.

Southern California Locoweed is common in Coastal Sage from the Palisades to Point Mugu. It blooms from February to April. We have three annual and two other perennial plants of this genus in our mountains; two of them are rare and the others occur in limited locations.

Many species of this genus are called Locoweed because animals exhibit bizarre behavior after eating the herbage. *Loco* means "crazy" in Spanish.

Astragalus is from Greek meaning "anklebone" and was applied long ago to some plants in this family because of the shape of the seeds. *Trichopodus* means "hairy pod." *Leucopsis* means "white."

Canyon Pea

Lathyrus laetiflorus Greene. ssp. *barbarae* (White) C.L. Hitchc.
Pea Family. FABACEAE (LEGUMINOSAE)

Linda Hardie-Scott

The Canyon Pea is a twining perennial. The compound leaves have 8 to 12 linear leaflets with tiny, sharp tips ending in a tendril. The 5 to 12 flowers appear on short stalks on either side of the main stem. They are pink, more-or-less streaked with lavender, and are ⅝ to ¾ inch long. The fruit is a many-seeded pod.

Canyon Pea blooms from March to May and is commonly found climbing over shrubs in Coastal Sage, Chaparral and Oak Woodland throughout.

The blossoms are instantly recognized as belonging to the Pea Family. They fade to tan and remain as later ones open, presenting a rather unattractive appearance. Pride of California *(L. splendens)* with showy, deep crimson flowers, is found in San Diego and Riverside Counties, but not, alas, in the Santa Monicas.

Lathyrus is the old Greek name for "pea." (Our cultivated Sweet Pea was developed from a Mediterranean species.) *Laetiflorus* means "abundantly flowering." *Barbarae* refers to Santa Barbara, the type locality.

Coastal Lotus

Lotus salsuginosus Greene.
Pea Family. FABACEAE (LEGUMINOSAE)

Tim Thomas

Coastal Lotus is a spreading annual with much-branched, leafy stems 4 to 12 inches long. The leaves are pinnately compound with 5 to 8 rounded leaflets covered with soft, fine hairs. The yellow flowers, under ½ inch long and subtended by one bract, are in 1 to 5-flowered clusters arising on stalks from a central point. The straight legumes are ½ to 1 inch long.

Coastal Lotus is found on sandy beaches and sea bluffs and in Coastal Sage and openings in Chaparral at low elevations, blooming from February to May.

the Mediterranean and North Africa. Two yellow-flowered species, Black Medic *(M. lupulina)* and Burclover *(M. polymorpha)*, are frequently encountered also.

The value of the herbage for feed has been known for centuries. The roots reach down 8 or 9 feet in the soil and once established the plant can yield several crops a year.

Medice is the Greek name for Alfalfa since it came to Greece from Medea. *Sativa* means "that which is sown."

White Sweet Clover
Melilotus albus. Desr.
Pea Family. FABACEAE (LEGUMINOSAE)

James P. Kenney

The White Sweet Clover is an erect, smooth-stemmed biennial from 3 to 6 feet high with thick, oblong, finely-toothed leaflets, in three's, rounded at the top. The pea-type, white flowers in slender one-sided groups are numerous. The fruit is an egg-shaped, oval pod.

White Sweet Clover and Yellow Sweet Clover *(M. indicus)* are common in waste places below 2000 feet throughout. They are both originally from Eurasia and both bloom from April to September.

If the breeze brings you a lovely smell reminiscent of new-mown hay when you are walking in the mountains, you may be sure you are near a mass of Sweet Clover. Even the leaves are fragrant. It is claimed that the flowers have been used for flavoring cheese, snuff and tobacco. Beekeepers recognize them as important honey plants.

Melilotus comes from two Greek words which mean "honey" and "a leguminous plant." *Albus* means "white."

Chaparral Pea
Pickeringia montana Nutt.
Pea Family. FABACEAE (LEGUMINOSAE)

Geoffrey Burleigh

The Chaparral Pea is an erect shrub 2 to 6 feet high with spiny branches. The few compound leaves have 3 leaflets and hug the stems. The solitary, typically pea-shaped, magenta flowers also hug the stems. There are 10 distinct stamens. The fruit is a 2-inch pod.

Chaparral Pea blooms in April and May above 1000 feet in Chaparral in the western part of the mountains.

This genus consists only of this species. The fruit rarely matures and the plant propagates from exposed root shoots. Where a thicket of this shrub has proliferated, penetration is impossible.

Charles Pickering (1805-1878) of the Philadelphia Academy of Sciences came to California in 1841 with the Wilkes Expedition as a physician and botanist. *Montana* means "mountains."

Leather Root
Psoralea macrostachya DC.
Pea Family. FABACEAE (LEGUMINOSAE)

Geoffrey Burleigh

The Leather Root is an erect perennial 1½ to 10 feet tall. The herbage has a strong odor and dark glands. The stalked leaves have 3 narrowly-oval leaflets and are subtended by conspicuous bracts. The 5 unequal sepals are covered with shaggy hairs and the lower lobe is longer than the pea-shaped, purple flowers which appear in spikes. The small pod is covered with tiny, matted hairs.

Leather Root is found infrequently in shade along canyon streams at low elevations, westward from Malibu Creek. It blooms from May to October.

P. physodes, the other member of this genus in the Santa Monica Mountains, is rarely encountered. It prefers dry places and has whitish flowers with only a slight purple touch.

Psoralea is from Greek meaning "roughly scaled" referring to the glandular dots on the leaves. *Macro* means "large" and *stachya* means "an ear of corn" and refers to the spikes of the inflorescence.

Spanish Broom
Spartium junceum L.
Pea Family. FABACEAE (LEGUMINOSAE)

James P. Kenney

The Spanish Broom is a tall, almost leafless shrub up to 9 feet tall. When leaves are present, they are small, ½ to 1 inch long, oval and smooth-margined. The yellow, pea-type flowers are large, up to an inch long, fragrant, and grow on short stalks on both sides of the main stem. The linear pod has many seeds.

This Mediterranean native has naturalized on dry slopes in the eastern half of the mountains and along road cuts elsewhere. It blooms from April to June.

Spanish Broom, an attractive nuisance, remains a showy yellow beauty when most natives have faded and folded. But it and its immigrant relatives, French Broom *(Cytisus monspessulanus)* and Scotch Broom *(C. scoparius)*, are aggressive spreaders, thriving so well that native plant enthusiasts have organized "broom bashes" to reduce their numbers and give the threatened local flora a fair chance.

Spartium is from a Greek word meaning "broom." The broom we use to sweep was originally made of Old World plants with this common name; thus it earned its name, broom. *Junceum* means "rush-like" and refers to the leafless stems.

Tomcat Clover
Trifolium tridentatum Lindl. var. *tridentatum*
Pea Family. FABACEAE (LEGUMINOSAE)

Linda Hardie-Scott

Tomcat Clover is a low herb 4 to 16 inches tall. As in all clovers, the compound leaves have 3 leaflets. The minutely-toothed leaflets are narrowly oblong, ½ to 1½ inches in length. At the base of each leaf stalk, there are prominent, narrow, leaf-like appendages. The flowers, typically pea-shaped, are pink-purple with a darker center, sometimes tipped with white. They are disposed in a broad, flattish head an inch or so across on footstalks 1 to 2 inches long. The fruit is a small 2-seeded pod.

Tomcat Clover is found occasionally on grassy slopes throughout, blooming from March to May.

We have 11 species of *Trifolium* in our mountains, two of which are common weeds. Most of those that are native are rarely seen now. They once formed a very important dietary staple of the California Indians who ate the tender leaves, both raw and cooked, as well as the seeds. Some tribes celebrated the spring appearance of the clover with special ceremonial dances.

Trifolium is from Latin meaning "3-leaved." *Tridentatum* means "3-toothed."

Spring Vetch
Vicia sativa L.
Pea Family. FABACEAE (LEGUMINOSAE)

Spring Vetch is an annual climbing plant, freely-branching with compound leaves consisting of about 7 pairs of leaflets and a terminal tendril. The pea-shaped flowers occur singly or in twos where the leaves join the stems. They are about an inch long on short stalks and reddish-purple in color. The fruit is a hairy pod 1½ to 3 inches long with 5 to 12 marbled seeds.

Spring Vetch is found at roadsides, in waste ground and pastures in the western half of the area blooming in April and May. It is probably a native of western Asia.

A similar, but smaller, black-podded species, *V. angustifolia,* can be found at Topanga and Malibu Creek. Another, *V. villosa,* covers large areas at Reagan Ranch. Many "Wild Sweet Peas" turn out to be *V. americana* which bears a close resemblance to *Lathyrus laetiflorus.*

Vicia is the classical Latin name for this genus. *Sativa* means "it is sown as a crop."

Gentian Family. GENTIANACEAE

Members of the Gentian Family are herbs with smooth-margined, opposite leaves. The flowers have parts in 4's or 5's. The stamens, inserted on the tube, are as many as the flower lobes and alternate with them. The ovary is superior and the fruit is a capsule with numerous seeds.

All parts of the plants of this family are very bitter and many species provide tonics and remedies all over the world.

The Gentians are the best known plants of the family, many of which are very showy and much admired.

We have only one species in our mountains, *Centaurium venustum.*

Canchalagua
Centaurium venustum (Gray) Rob.
Gentian Family. GENTIANACEAE

Linda Hardie-Scott

Canchalagua is an erect annual 6 inches to 2 feet high. The pale green, opposite leaves arise directly from the main stem. The 5 bright pink petals have red spots on their white throats. The cup-shaped sepals are 5-parted and the 5 stamens have anthers that are spirally twisted after they have shed their pollen. (This exquisite spiraling of the anthers rewards a look with a hand lens.) The stigma is fan-shaped. The fruit is a capsule with numerous seeds.

Our only little Gentian blooms in June and July in drying grasses in open areas of Chaparral and Coastal Sage in the central parts of the range.

This was considered by the early Spanish settlers to be an indispensable remedy for fevers. Bundles of dried Canchalagua hung from the kitchen rafters of every *hacienda.*

The generic name comes from the Latin *Centaur* who was supposed to have discovered its medicinal uses. *Venustum* means "handsome, charming" which this bright little wildflower certainly is.

Geranium Family. GERANIACEAE

The members of this family are herbs with lobed or compound leaves. The radially symmetrical flowers have 5 sepals, 5 petals and stamens in circles of 5, 10 or 15. The family is especially characterized by the long beak on the fruit—like the beak on a crane. The mature fruit splits into parts which then curl up spilling their seeds.

In cultivation geraniums and pelargoniums are well regarded for their large colorful flowers.

Geranium is derived from the Greek word for "crane or heron."

Red-stem Filaree
Erodium cicutarium (L.) L'Hér.
Geranium Family. GERANIACEAE

James P. Kenney

Red-stem Filaree is an annual or biennial herb with stems 3 to 12 inches long. The hairy, dark green, finely dissected leaves are ½ to 4 inches long and form a rosette close to the ground when the plant is young. Slender footstalks topped with magenta ¼ inch flowers rise from a single point at the top of the stem. The fruit divides into 5 one-seeded parts with a twisted, hairy tail that coils when moistened.

Red-stemmed Filaree is a very common weed in open waste places and grassy slopes. It blooms from January until May.

This is by far the most widespread of all our *Erodium* species and is highly regarded as forage for all classes of livestock. It is believed to be an introduction from the Old World but a very early one as Fremont reported it in 1844 covering the ground "like a sward." You will also find Broad-leaf Filaree *(E. botrys)* which has larger wine-veined pink flowers and White-stem Filaree *(E. moschatum)* which has larger leaves, toothed but not cleft, and whitish stems.

Filaree is a corruption of the Spanish name *Alfilerilla,* "a needle," another interpretation of that beak.

Erodium comes from Greek and means "heron's bill." *Cicutarium* refers to the leaves which resemble the leaves of Poison Hemlock. *Cicuta* is the ancient Latin name for Poison Hemlock.

Carolina Geranium
Geranium carolinianum L.
Geranium Family. GERANIACEAE

Carolina Geranium is an annual 8 to 16 inches high. The foliage is covered with soft, short hairs, the leaves palmately 5 to 7-parted, and the segments lobed. The petals about ¾ inch long are pinkish and the sepals have a short bristle at the tip. The covering of the seed is hairy and usually black.

Carolina Geranium prefers shady locations and is scattered throughout at low elevations. It blooms from March to June.

G. carolinianum appears to be a native in California and as you might well expect from its names is also from the eastern part of the United States. *G. molle* introduced from Eurasia and North Africa is taller, has rose-purple petals, but no hairs on the seed coverings. It is found near gardens in the vicinity of West Los Angeles and Santa Monica.

Geranium means "crane," from the beak-like fruit; *carolinianum* means "from Carolina."

Waterleaf Family. HYDROPHYLLACEAE

Waterleaf Family plants are mostly herbs, a few shrubby, with mainly alternate leaves. The tubular flowers are 5-parted and are often on coiled stems. The 5 stamens alternate with the lobes of the flower tube. There are 2 styles, a superior ovary and the fruit is a capsule.

Most of the members of this family are North American and are concentrated in the West. In the Santa Monicas we have 7 genera and 21 species, all of them natives.

Whispering Bells

Emmenanthe penduliflora Benth.
Waterleaf Family. HYDROPHYLLACEAE

Linda Hardie-Scott

Whispering Bells are yellow-green, hairy annuals 6 to 12 inches high. The narrow leaves are an inch or more long with numerous, shallow-toothed lobes. The creamy or light yellow, bell-shaped flowers are erect at first, but eventually droop on threadlike stalks.

Whispering Bells bloom in April and May in Chaparral and Oak Woodland throughout and are more abundant after fires.

The freshly opened flowers do resemble bells but they do not whisper. Found late in the season, however, they are dry as paper and rustle in every passing breeze so the common name is most appropriate. So consistently do they remain on their stems they may be used for dry decorations.

The generic name is a way of saying in Greek "the flower that abides," an allusion to the fact that the blossom does not fall as it fades. *Penduliflora* means that the blossom "hangs down," as it does in age.

Thickleaf Yerba Santa

Eriodictyon crassifolium Benth.
Waterleaf Family. HYDROPHYLLACEAE

Steven R. Kutcher

Thickleaf Yerba Santa is an erect shrub 3 to 12 feet high with dense, short, matted hairs on its leaves, stems and sepals. The alternate, gray-green, oval leaves can be 6 inches long and 2 inches wide. The pale blue-violet flowers, about ½ inch long, have a tubular base and an expanded upper part. They occur in one-sided branches. The fruit is a white, gummy capsule.

Thickleaf Yerba Santa may be found in Chaparral and Coastal Sage on the northern slope of the western portion of the range. It blooms in April and May.

It is reported that the Spanish padres gave this plant its common name of Yerba Santa, "Holy Herb," for they as well as the Indians found it a famous remedy for respiratory infections and fevers. The Indians bound the fresh leaves on sores of man and animal and even smoked and chewed the dried ones.

When a fresh leaf is picked and chewed on the trail, there is first a bitter taste which changes to a sweet, cool one.

Eriodictyon is derived from two Greek words meaning "hairy net" referring to the underside of the leaves. *Crassifolium* means "thick-leaved."

Common Eucrypta

Eucrypta chrysanthemifolia (Benth.) Greene.
Waterleaf Family. HYDROPHYLLACEAE

Margaret Stassforth

Common Eucrypta is an erect, hairy annual 8 to 20 inches high with leaves dissected into rounded lobes. The lower leaves are opposite and the upper ones are alternate. The tiny, white flowers are bell-shaped and appear in terminal clusters.

Common Eucrypta is found throughout the range away from the immediate coast. It blooms from February to May.

This plant is particularly abundant and very noticeable after fires when it densely covers large areas.

Eucrypta is from two Greek words meaning "true secret," referring to the hidden inner seeds. *Chrysanthemifolia* was given because the foliage resembles that of the chrysanthemums.

Baby Blue Eyes

Nemophila menziesii H. & A.
Waterleaf Family. HYDROPHYLLACEAE

Geoffrey Burleigh

Baby Blue Eyes are low, rather hairy annuals spreading by many tender stems 4 to 12 inches long. The opposite leaves are divided into 7 or 9 parts which may be lobed again. The wheel-shaped flowers have a lighter center often with black or dark blue dots. They can be as large as an inch across.

This delightful little wildflower blooms from February to June in grassy areas under oaks or in Chaparral in the eastern part of the range.

David Douglas, a Scottish botanist who visited California in 1831 and 1832, introduced this native to Europe where it has long been a garden favorite. One authority claims Douglas collected 500 species in the state and added more to the knowledge of California botany than all who had gone before him.

Baby Blue Eyes are easily grown from seed broadcast in fall. They are water tolerant, like sun to a bit of high shade and are not particular about soil. They serve especially well as a ground cover in a bulb garden.

Nemophila means that it has "an affinity for groves." The species is named for Archibald Menzies (1754-1842), a Scottish botanist who combined his knowledge of plants with his skill as a surgeon on an expedition (1790-1795) with Captain George Vancouver (1758-1798) on the *Discovery*. They visited California in 1792-1794.

Caterpillar Phacelia

Phacelia cicutaria Greene var. *hispida* (Gray) J.T. Howell.
Waterleaf Family. HYDROPHYLLACEAE

Barry Silver

The Caterpillar Phacelia is an annual with stiff hairs on its stems and leaves. The leaves are pinnately compound with toothed lobes. The dirty-white to pale lavender flowers on coiled stems are also covered with stiff hairs.

P. cicutaria occurs most commonly in Chaparral, frequently on burns throughout the mountains.

The tight, hairy coils of the flowering stems of this particular Phacelia have given it the common name.

The generic name is based on a Greek word meaning "cluster" alluding to the densely crowded flower spikes of most species of this genus. The very first species was collected almost 200 years ago in the Straits of Magellan. *Cicutaria* refers to *Erodium cicutarium* whose leaves were thought to resemble *cicuta.*

Common Phacelia

Phacelia distans Benth.
Waterleaf Family. HYDROPHYLLACEAE

Linda Hardie-Scott

The Common Phacelia is a straggling, much-branched annual with hairy leaves and stems 1 to 2 feet high. The leaves are finely dissected, and a few inches long. The ½ inch, blue to whitish, round flowers with stamens only slightly exserted occur in clustered coils that unroll gradually as the flowers expand.

Common Phacelia occurs below 2000 feet in Oak Woodland, Coastal Sage and on ocean bluffs throughout the mountains. It blooms from March to June.

Very similar to *P. distans* and often mistaken for it is the Tansy-leaved Phacelia *(P. tanacetifolia),* whose stamens extrude much further. The early Spanish called both of them *Vervenia,* a possible diminutive of Verbena.

Distans means "separated, apart" in reference to the long, exserted stamens which are apart from each other.

Large-flowered Phacelia
Phacelia grandiflora (Benth.) Gray.
Waterleaf Family. HYDROPHYLLACEAE

Linda Hardie-Scott

The Large-flowered Phacelia is a robust plant 1 to 3 feet high covered with sticky hairs. The oval leaves are coarsely toothed and 2 to 3 inches long. The deep lavender flowers streaked with purple are saucer-shaped and showy. They may be up to 2 inches across. The stamens are long and purple and the anthers are large.

This spectacular Phacelia blooms from February to June in Chaparral, Oak Woodland and Coastal Sage throughout. It is especially abundant after fires.

All of the Phacelias are native to the New World. Authorities claim 87 of them for California, 13 of which are in the Santa Monicas. Although casual identification of many of them is very difficult, this one is certain since it has larger flowers than any other. The sticky glands covering the whole plant will turn everything they contact to a deep red-brown resembling iron rust. This stain can never be removed completely from clothing and stays on skin almost as persistently. The hairs carry a small globule of oil that can cause a blistering rash on the skin of sensitive people.

Grandiflora means "large-flowered."

Sarah Thomas Schwaegler

Imbricate Phacelia

Phacelia imbricata Greene.
Waterleaf Family. HYDROPHYLLACEAE

Geoffrey Burleigh

The Imbricate Phacelia is a greenish-gray perennial 8 to 16 inches high densely covered with hairs. The oval, alternate leaves are mostly basal and pinnately divided into 5 to 9 leaflets with pronounced parallel veins. The tubular, white flowers, ¼ inch long, occur in dense coils.

P. imbricata is found inland in the shaded margins of Oak Woodland in the western half of the mountains blooming in May and June.

All of the Phacelias are easily grown from seed which should be scattered in the autumn. They prefer sun and, unlike many natives, are water tolerant.

Imbricata means "overlapping, closely put together" referring to the calyx lobes which are imbricate laterally in fruit.

Parry's Phacelia

Phacelia parryi Torr.
Waterleaf Family. HYDROPHYLLACEAE

Dede Gilman

Parry's Phacelia is an erect annual up to 2 feet tall. The brittle stems are gummy and covered with stiff hairs. The oval, toothed leaves are ½ to 2 inches long. The wheel-shaped flowers are dark purple-blue with paler centers and are ⅜ to ¾ inch long. The 5 stamens stand well above the petals.

Parry's Phacelia blooms from March to May and is abundant in disturbed areas of Chaparral and Coastal Sage in the western half of the mountains throughout. It is especially plentiful after fires.

P. parryi has a very beautiful flower with rich color and alert stamens. Wild Canterbury Bell *(P. minor),* less frequently encountered, is very similar, but the flowers are more tubular.

Dr. C. C. Parry (1823-1890) was an American botanist who visited the southwestern mountains and deserts many times. He was once connected with the Mexican Boundary Survey and is remembered in the names of more than a score of native plants.

Branching Phacelia
Phacelia ramosissima Dougl. ex Lehm.
Waterleaf Family. HYDROPHYLLACEAE

This is a straggling, branching perennial with rough, hairy, pinnately-divided leaves and main stems up to 3 feet long. The flowers are clustered and coiled at the ends of the stems. Near the coast they are dark blue. Inland they are dirty white.

The inland variety, *P. suffrutescens,* may be found throughout the mountains blooming from April to September.

This species, often going under the name Wild Heliotrope, clambers up through other shrubs or makes great mounds of foliage along streams and embankments. It is an important honey plant.

Ramosissima means "very much branched."

Sticky Phacelia
Phacelia viscida (Benth.) Torr.
Waterleaf Family. HYDROPHYLLACEAE

Tim Thomas

The Sticky Phacelia is an annual, 1 to 2 feet tall, branching from the base with hairy, roundish, slightly toothed leaves. The few to many deep blue flowers are also round and have a whitish center. They may be ¾ inch across and appear on either side of the stem at the top.

Sticky Phacelia may be found in disturbed Chaparral and Oak Woodland from the middle to the eastern parts of the mountains. It blooms from March to May.

Like *P. grandiflora,* this plant has many viscid glands and will stain hands and clothing on contact.

Viscida means "sticky."

Fiesta Flower
Pholistoma auritum (Lindl.) Lilja.
Waterleaf Family. HYDROPHYLLACEAE

Linda Hardie-Scott

The Fiesta Flower is a trailing annual. The square stems are 1 to 3 feet long and quite weak with back-pointing prickles. The leaves are pinnately divided into 5 to 9 lobes and have a clasping, eared base. The violet petals are fused at the bottom, have 5 lobes and are an inch or so across. Each of the 5 sepals has a turned-down appendage.

Fiesta Flower blooms from March to May in brushy areas of Chaparral and Oak Woodland. It is most frequent after fires.

It is said that the señoritas of early days decorated their ball gowns with sprays of Fiesta Flower which will cling tightly to thin fabrics. It is pleasant to think of guitars plinking softly on a moonlit hacienda patio as a dark-eyed young lady stepped out in the fandango, graced with these lovely garlands. It makes the common name appropriate.

Pholistoma means "scaly-mouthed" and was given because of the scales in the throat of the flower. *Auritum,* "eared," refers to the clasping, eared base of the leaves.

Mint Family. LAMIACEAE (LABIATAE)

The Mint Family has herbs and shrubs with square stems and opposite leaves. All of our native mints are aromatic. The flowers are 2-lipped, the upper lip with 2 lobes and the lower lip with 3. The stamens are typically 4 with 2 shorter than the others. There is a single, 2-parted style and a superior ovary. The fruits are 4 one-seeded nutlets.

There are no poisonous plants in this family. Many are cultivated for their beautiful flowers and even more for the condiments they provide, such as sage, thyme, peppermint and marjoram.

The Santa Monica Mountains have 10 genera and 20 species of this family.

Horehound
Marrubium vulgare L.
Mint Family. LAMIACEAE (LABIATAE)

James P. Kenney

Horehound is a perennial herb in clumps a foot or 2 high with square, woolly, white stems, but no mint odor. The roundish leaves are gray-green with prominent veins and wrinkles and covered beneath with matted, white hairs. The small white flowers are crowded in dense clusters in the axils of the upper leaves. The calyx has 10 short teeth which turn back and become hooked at the tips. The fruit becomes a bur when dry and attaches itself to clothing and animal fur spilling the little nutlets where they will prosper.

Horehound is a common weed in waste places and old fields. It blooms in spring and summer.

This is a European species which has become widespread throughout the United States. It probably came into California with the Americans who brought a root or two to plant as a remedy for colds. Horehound candy sucked for a sore throat used to be available in drug stores. It can be made by boiling a cup of Horehound leaves in 2 cups of water for 10 minutes. Add twice as much sugar or honey and a pinch of cream of tartar and continue to boil until the hard crack stage, or 290 degrees. Pour on a buttered plate and you will have an old fashioned, cough drop candy. If a little lemon juice is added when boiling, it will taste less bitter.

Marrubium is from a Hebrew word meaning "bitter." *Vulgare* means "commonplace,"—which this weed certainly is.

Mustang Mint
Monardella lanceolata Gray.
Mint Family. LAMIACEAE (LABIATAE)

Dede Gilman

The Mustang Mint is a branching, aromatic annual from 6 inches to 2 feet high. The few narrow leaves taper at both ends and are 1 to 2 inches long. The small, blue-purple flowers with purplish bracts underneath mass in crowded terminal heads over an inch across.

Mustang Mint occurs in oak woodlands near Lake Sherwood and Stokes Canyon.

It is frequently found in other parts of the state, but not often in the Santa Monicas. It blooms in May and June.

There is a pungent fragrance to this plant. The Spanish-Californians called it *Poleo* and used its leaves for remedies and teas. *Poleo* is the name of a Pennyroyal *(Mentha pulegium)* in Spain which repels fleas. One stem of *M. lanceolata* complete with flowers makes two cups of tea. However, Mustang Mint is neither abundant nor weedy enough to be gathered for tea today.

Nicolás Bautista Monardes (1493-1588) was a 16th century Spanish physician and botanist whose name was given to a number of plants. He studied American drugs at the docks in Seville, had a large investment in importing drugs from the Americas as a business and was one of the first to test drugs on animals. He had a large medical practice and was the best known and most widely read Spanish physician in Europe in his time. He was an expert botanist. His book, the first to be published on the flora of the Americas, was translated into English under the title, *Joyfull newes out of the new founde worlde.*

Monardella is a diminutive. *Lanceolata* means "lance-like" and refers to the narrow leaves.

White Sage
Salvia apiana Jeps.
Mint Family. LAMIACEAE (LABIATAE)

Dede Gilman

White Sage is a shrub 3 to 6 feet tall with erect branches. The crowded, opposite, pointed leaves have small, rounded teeth on the margins. They are such a light gray as to appear almost white. The pale lavender to white flowers, ½ to ¾ inch long with stamens well exserted, do not appear in separated whorls as in other members of this genus but in loose heads on either side of the stem.

White Sage is most often encountered in Coastal Sage but never on the immediate coast. It can also appear in Chaparral and Oak Woodland throughout. It blooms from April to June.

This plant well deserves its species name which refers to bees. The honey made from it is clear, pale and superfine.

The blossoms have especially adapted themselves to pollination. The alighting bee first brushes against the stigma and leaves pollen from another flower that it had recently visited. When it sits on the lower lip to get at the nectar, it bends down and grasps the 2 stamens as handles which draws the anthers to its body leaving pollen to take to the next blossom.

The *Salvias,* or Sages, as they are often called, take their name from a Latin herb, *Salvia,* used for healing. The name comes from the Latin *salveo,* "I am well." *Apiana* pertains to bees which this plant attracts in great numbers.

Chia
Salvia columbariae Benth.
Mint Family. LAMIACEAE (LABIATAE)

Dede Gilman

Chia is a purplish-stemmed annual from 4 inches to 2 feet high. The dull green, coarsely-wrinkled leaves, 1 to several inches long, are mostly at the base. They are deeply cut into toothed, blunt lobes. The small, blue flowers, barely exceeding the spiny-toothed sepals, are markedly 2-lipped, the upper lip with 2 lobes and lower one drooping with 3 lobes. The flowers appear in interrupted whorls underneath which are many wine-colored, prickly bracts.

Chia blooms in April and May on open banks mostly in Coastal Sage away from the immediate coast.

The tiny Chia seeds, rich in mucilage and oil, are famous from very ancient times, a staple food of the Pacific Coast and Mexican Indians, even having been a cultivated crop of the latter. The original "Chia" of the Aztecs was *Salvia hispanica.* The Indian method of harvesting the seeds was to beat the heads with a paddle over a flat basket. A single teaspoon of seed was reported to have been able to sustain a man on a 24-hour, forced march. A seed or two was placed under the eye to alleviate eyestrain. The plant was also used to neutralize alkaline water.

These seeds may be bought today in health food stores. A refreshing drink which assuages great thirst may be made from the crushed, parched seeds steeped in water with the addition of lemon and sugar.

This species reminded its namer of *Columbaria,* a plant in the Old World genus of *Scabiosa.*

Purple Sage
Salvia leucophylla Greene.
Mint Family. LAMIACEAE (LABIATAE)

Dede Gilman

The Purple Sage is a shrub 3 to 5 feet tall, with gray, hairy herbage. The leaves, ¾ to 2 inches long, are longer than wide and have small rounded teeth on their edges. They are heart-shaped at the base and have prominent veins. The rosy-lavender flowers are in compact whorls with gray, oval, leaf-like bracts underneath.

Purple Sage is common in Coastal Sage throughout the area blooming from May to July.

Both the stems and the bracts below the flowers feel mealy to the touch. The large spikes of rich, warm, lilac flowers are a delight to see when encountered on a covered hillside. Like the other *Salvias,* it is favored by bees and may hybridize with other species.

Leucophylla means "white-leaved."

Black Sage
Salvia mellifera Greene.
Mint Family. LAMIACEAE (LABIATAE)

Geoffrey Burleigh

The Black Sage is a shrubby plant 3 to 6 feet high with narrowish, wrinkled, dark green leaves which when crushed have a strong, minty odor. The light blue flowers are about ½ inch long, deeply 2-lipped and grow in 3 to 9 dense whorls along the slender stems.

Black Sage may be found throughout the area blooming from March to June.

Entire sunny hillsides of Chaparral are sometimes covered with thickets of Black Sage humming with nectar-gathering bees. Indians gathered the tiny seeds, parched them and made them into meal.

Black Sage became a common name because the whorls of blooms remain after they set seed forming dark spheres along the dry stalks.

Mellifera means "honey bearing."

Hummingbird Sage, Pitcher Sage
Salvia spathacea Greene.
Mint Family. LAMIACEAE (LABIATAE)

Linda Hardie-Scott

The Hummingbird Sage is a coarse perennial with woolly stems, 1 to 2½ feet high. The opposite, wrinkly leaves have rounded teeth on the margins and are white-woolly beneath. They are oblong to arrow-shaped at the base and 3 to 8 inches long. The stalk measures over a foot tall with many large, widely separated whorls of 2-lipped crimson flowers, 1 to 1½ inches long with the stamens projecting beyond them. The oval bracts underneath the whorls and the stems are purplish-bronze colored, a pleasing contrast to the dark, glowing red of the trumpet-shaped blossoms.

S. spathacea blooms in partial shade from March to May in Oak Woodland and Chaparral throughout the range but away from the immediate coast.

Pitcher Sage produces much nectar and hummingbirds seem to be the major beneficiary.

Spathacea meaning "with a spathe" refers to the large, colored bract that encloses the flower cluster.

Danny's Skullcap
Scutellaria tuberosa Benth.
Mint Family. LAMIACEAE (LABIATAE)

Dede Gilman

Danny's Skullcap is a low, erect herb 1 to 8 inches high. The opposite leaves, about ½ inch long, are oval with round, toothed margins. The 2-lipped flowers are dark purple, ½ to ¾ inch long and grow in groups of 1 to 3 in the angle between the leaf and the stem. The sepals form a 2-lipped tube. The 4 stamens contained in the helmet are arranged in pairs curving upward.

Skullcap blooms from March to May in Chaparral and Oak Woodland from Boney Ridge east away from the coast.

Notice how, if there are 2 tiny blossoms

arising from the point where the leaves meet the stems, they twist so that they stand side by side in pairs. The unopened sepals have been compared to the shape of an old-fashioned Quaker bonnet. Quaker Bonnets might have been a more appropriate common name than Skullcap.

Scutellaria is from Latin and means "a tray or platter." This refers to the sepals which appear this way during the fruiting period. *Tuberosa* means "tuberous." The rhizomes of this plant have tubers.

California Hedge Nettle

Stachys bullata Benth.
Mint Family. LAMIACEAE (LABIATAE)

Geoffrey Burleigh

The California Hedge Nettle is a slender, hairy, weak-stemmed herb 16 to 32 inches high. The opposite, oval leaves are 1 to 7 inches long, round-toothed with a slightly heart-shaped base. The lavender to red-violet flowers with white markings on the lower lip appear in whorls and are ½ to ¾ inch long. The sepals are only ¼ inch long and are spiny.

California Hedge Nettle blooms from March to May in moist areas below 2000 feet in the western half of the range.

The common name is misleading. The true nettles, *Urtica*, have stinging hairs which continue to burn the skin of those unfortunate enough to touch them for hours. This plant is not punishing. In fact, the leaves have a very pleasant lemon scent when rubbed.

Stachys comes from the Greek word that means "ear of grain," hence a spike, the growth form the flowers take. *Bullata* means "blistered, puckered," describing the texture of the leaves.

Woolly Blue Curls

Trichostema lanatum Benth.
Mint Family. LAMIACEAE (LABIATAE)

Claire Kaye

The blue flowers are nearly an inch long and densely clothed with violet wool. The buds and the sepals are pinkish-purple. The 4 stamens are conspicuous, curved and hang far out. The stems are square and the opposite leaves are thick, narrow, turned under at the edges, shiny above and white-woolly beneath. It is a perennial shrub 2 to 5 feet high.

Woolly Blue Curls bloom in Chaparral and Coastal Sage from Mandeville Canyon westward from March to June.

This showy plant is sometimes called *Romero,* Spanish for "Rosemary," which its foliage does resemble. The early Californians made a liniment of the leaves for bruises.

Trichostema means that the stamens are hair-like. *Lanatum* means "woolly."

Stick-leaf Family. LOASACEAE

These are erect herbaceous plants with rough, barbed leaves, often with stinging hairs. The single flowers on small stalks appear on either side of the stems and have 5 or 10 petals. The ovary is inferior and the stamens are numerous.

The members of this family are chiefly American. There are over two hundred species in some fourteen genera although the one described here is one of two found in the Santa Monicas.

Small-flowered Stick-leaf

Mentzelia micrantha (Hook. & Arn.) T. & G.
Mint Family. LAMIACEAE (LABIATAE)

Linda Hardie-Scott

The Small-flowered Stick-leaf is an erect annual 8 to 20 inches tall with rough, grayish stems. The rough leaves are toothed or lobed, 1 to 3 inches long. The tiny, pale yellow flowers, often concealed by leafy bracts, are clustered at the tips of the branches. There are 5 petals, 5 sepals and many stamens. The fruit is a small cylindrical capsule.

M. micrantha occurs on sandy banks and dry, open, disturbed places throughout. It blooms in April and May.

There are a half dozen or more species of *Mentzelia* on the Pacific Coast, most of them larger and far more showy than this one. They are easily recognized by the conspicuous clustered stamens and the top-shaped calyx with its 5 noticeable lobes. The barbed hairs covering the plants cause them to cling to clothing which gives rise to the common name. Leaves that adhered to jeans, tennis shoes or canvas bags can still be recognized after several trips to the laundromat.

The generic name honors Christian Mentzel (1622-1701), a 17th century German botanist, botanical author, philologist and physician—and father of the first king of Prussia. *Micrantha* combines two Greek words meaning "small-flowered."

Mallow Family. MALVACEAE

Members of the Mallow Family are herbs or shrubs often with showy flowers. The flowers have 5 petals. The stamens are fused into a center column surrounding the style. The fruit varies.

Marsh Mallow *(Althea officinalis)*, a member of this family, once provided a gluey substance from its roots to make marshmallows. This confection is made now from a mixture of corn syrup, sugar, albumen and gelatin.

The Hibiscus and the Hollyhock found in gardens are plants from this family.

We have only 3 native species in the Santa Monica Mountains.

Bush Mallow

Malacothamnus fasciculatus (Nutt.) Greene.
Mallow Family. MALVACEAE

Dede Gilman

The Bush Mallow is a shrub 3 to 15 feet tall with wand-like branches. The gray-green, densely-hairy leaves have 3 to 5 lobes. The pale pink-lavender flowers have 5 petals and are arranged in spikes. The numerous stamens with golden brown anthers are fused into a tube around the style. The flowers have a delicate perfume. The fruit is a schizocarp, which at maturity breaks into several parts, each containing a seed.

Bush Mallow blooms from April to October and is abundant in disturbed ground along roads, especially in Coastal Sage.

It is a handsome shrub when in full bloom and has sometimes been called False Mallow.

The generic name, *Malacothamnus*, comes from Greek and means "soft shrub." The species name, *fasciculatus*, refers to the clustered position of the leaves.

Cheeseweed
Malva parviflora L.
Mallow Family. MALVACEAE

Cheeseweed is a coarse annual 1½ to 3 feet tall. The roundish leaves are 1 to 5 inches broad, have 6 or 7 lobes, long leaf stalks and a red spot at the base of the blade. The 5-petaled flowers, resembling a small hibiscus, are only ¼ inch long. They are white to pink or bluish and cluster in the axils of the leaves where they often go unnoticed. The fruit is green, circular and flattened, separating on maturity into many one-seeded segments.

Cheeseweed is a native of the Mediterranean region. It blooms the year around and is a very common weed of waste places at low elevations throughout.

The resemblance of the fruit to a small round cheese gives the plant its common name. The slightly fuzzy leaves are not appealing when eaten raw, but when lightly boiled or steamed are considered tasty in any recipe utilizing cooked greens. Remove the husks from the seeds and use the seeds in salads. When chickens eat Cheeseweed they produce eggs with pink whites.

Malva is a Latin name from Greek, *malache,* referring to the leaves which were soothing to the skin. *Parviflora* means "small-flowered."

Checker Bloom
Sidalcea malvaeflora (DC.) Gray ex Benth.
Mallow Family. MALVACEAE

Dede Gilman

Checker Bloom is a perennial herb with stems 8 inches to 2 feet long. The basal leaves are round with shallow lobes; the stem leaves are cut into as many as 7 fingers with 3 lobed divisions. The pink or rose-purple flowers are in terminal clusters on small stalks on both sides of the main stem. The 5 separate petals are ½ inch to an inch long. There ae 5 sepals as well, and the stamens are united in a 2-series column.

You may look for Checker Bloom to show its flowers from March to July on north slopes of the mountains west of Malibu Creek.

Some call this plant Wild Hollyhock and they are not far off. Hollyhocks, though of a different genus, are of the same family. The Sidalcea, a native genus, is confined to the West and has more than a dozen species. This is the most common.

Sidalcea is a combination of two genera in the Mallow Family, *Sida* and *Alcea*. *Malvaeflorus* means "mallow-flowered."

Four O'Clock Family. NYCTAGINACEAE

The Four O'Clock Family consists of more or less succulent herbs or low shrubs with delicate stems and bulbous joints. The leaves are opposite and are not toothed or divided. There are no petals. The brightly-colored bell-shape is really a set of sepals.

The brilliance of our garden Bougainvillea, also in this family, must be attributed to sepals and bracts, not petals. It is probably best to enjoy the colorfulness of all of the Four O'Clocks without even trying to figure out the difference between petals, sepals and bracts as they appear in NYCTAGINACEAE.

We have two species of *Abronia* and one of *Mirabilis* in the Santa Monicas.

Beach Sand Verbena

Abronia umbellata Lam.
Four O'Clock Family. NYCTAGINACEAE

Linda Hardie-Scott

The Beach Sand Verbena is a perennial with sticky, hairy stems and oval, opposite leaves about 2 inches long. The stems trail on the sand. Only the branches that bear the rose-pink flowers stand erect. The flowers have white centers and are night-fragrant. They are ½ inch long and the thin tubular base widens into 5 lobes, each twice-cleft. They are clustered into a head. Small, white moths visit them at night, and Skippers, Hover Flies and small beetles by day. The fruit is a winged achene.

There are other species of *Abronia,* some on deserts, others in the mountains. Red Sand Verbena *(A. maritima)* may also be found on our beaches. Its stems and leaves are more succulent and its flowers are smaller and darker. The common name Sand Verbena has been given to the genus because the flower heads have a striking resemblance to the true Verbenas of the garden, although they are not in the same family. The *Abronia* species delight in growing in the sand.

Both species of Sand Verbena bloom from March until November. *A. umbellata* is found on sandy beaches all along the coast. *A. maritima* is located from Pt. Dume west.

Abronia is from Greek and means "graceful." *Umbellata* refers to the arrangement of the flowers which arises in a head from a central point. (Your umbrella has a similar derivation.)

Wishbone Bush
Mirabilis californica Gray.
Four O'Clock Family. NYCTAGINACEAE

Linda Hardie-Scott

The Wishbone Bush is a low, bushy perennial with fragile stems, woody below but herbaceous above, forking repeatedly and sometimes supporting themselves on nearby bushes. The opposite, smooth-margined leaves are slightly heart-shaped. The plant is softly hairy and sticky. There is a green cup supporting each flower. The 5 red-violet lobes, each 2-parted at the end, are not petals, but are delicate sepals. (This can be very confusing to the amateur botanist.) There are 5 stamens and a round ovary.

Wishbone Bush is pollinated at night. The blossoms open about mid-afternoon and do not close until early the next day, so it is easy to see how Four O'Clock, the family name, came about. Wishbone, as a common name for this species, is even easier to understand if you discover a dead plant. The dry, forked stems closely resemble the turkey or chicken bones we pulled and wished on as children.

Wishbone Bush blooms from March to May on rocky slopes at low elevations most often in Coastal Sage.

Mirabilis is Latin meaning "miraculous."

Evening Primrose Family. ONAGRACEAE

The Evening Primrose Family consists of herbs usually with simple leaves. The flowers have 4 petals and most commonly, 8 stamens. The stigma is either ball-shaped or 4-lobed. The ovary is inferior.

The name is misleading, for these plants are not related to the Primroses (PRIMULACEAE). The Primrose name was given to them in the early 1600's by John Parkinson, an English herbalist who was the first to describe these New World flowers. Their scent reminded him of the Wild Primroses in English meadows. The "Evening" of the name refers to the habit of some species of opening at sundown and closing by morning.

Our two species of *Oenothera* are lovely, but rarely seen. Our two species of *Epilobium* have minute flowers. The *Boisduvalia* once occurring in our flora is now extinct. Nevertheless, with the four genera we have, this is an important family in our native flora:

Southern Sun Cups

Camissonia bistorta (Nutt. ex T. & G.) Raven.
Evening Primose Family. ONAGRACEAE

Linda Hardie-Scott

The Southern Sun Cups are prostrate annuals with reddish stems radiating outward, usually 2 or 3 inches high. The leaves are about 3 inches long and somewhat toothed. The bright yellow flowers, often with a brown spot at the base of each petal, are nearly an inch across and are in the axils of the leafy stems. In the genus *Camissonia,* the stigma is round while that of *Oenothera* is 4-parted, a sure clue to separate these two very similar genera. As in all members of this family, the ovary is inferior and what appears to be a stalk under each blossom contains the seed chamber.

The Southern Sun Cups are common on sandy flats near the beach and scattered in grassy openings in Chaparral. They bloom from March to June.

They grow well from seed in sandy soil with full sun and moderate water.

Camissonia was named in honor of Adelbert Ludwig von Chamisso (1781-1838) who named the California Poppy. He was the botanist on the ship *Rurik* which visited California in 1816. *Bistorta* means "twice-twisted." The fruit has this double turn.

Beach Primrose

Camissonia cheiranthifolia (Hornem. ex Spreng.) Raimann ssp. *suffruticosa* (Wats.) Raven.
Evening Primrose Family. ONAGRACEAE

Barry Silver

The Beach Primrose is a perennial with prostrate stems and oval, smooth-margined leaves, the upper clasping. The stems and leaves are gray-green and covered with densely matted, silvery hairs. The bright yellow flowers are in the leaf axils and have 4 petals, ½ to an inch long. The sepals are bent backward. The capsules are 4-angled, covered with hairs and coiled when mature.

Beach Primrose is common along the entire coast and blooms from March until July.

This attractive plant forms dense mats above the tide line on sandy beaches and serves as an important dune stabilizer when undisturbed.

The specific name means that the leaves are like that of the *Cheiranthus,* an old name for a Wallflower now renamed. *Suffruticosa* leads to the English word "suffruticose" which means "very low, barely woody and shrub-like."

Farewell-to-spring
Clarkia deflexa (Jeps.) Lewis & Lewis.
Evening Primrose Family. ONAGRACEAE

Linda Hardie-Scott

Farewell-to-spring is an annual herb 1 to 3 feet high. The narrow, smooth-margined leaves are 1 to 3 inches long. The buds point down, rising to an erect position as they open. The reddish sepals are up to ¾ inch long, united and turned to one side. The 4, showy pink or lavender, fan-shaped petals are usually whitish toward the tips, dark at the base and flecked with violet. The capsule is 1 to 1½ inches long.

This *Clarkia* blooms from April to July on open slopes below 2000 feet throughout the mountains.

Farewell-to-spring is a name applied to many *Clarkia* species, although in Southern California, spring is by no means departing when they appear. *C. cylindrica* is very similar but its buds are erect. *C. purpurea* has a much smaller flower and is dark purple. Older floras list the genus as *Godetia.*

William Clark (1770-1838) of Lewis and Clark fame is honored by having his name attached to all of these lovely flowers. *Deflexa* means "bent, or turned abruptly downward at a sharp angle" as the buds and sepals are.

Elegant Clarkia
Clarkia unguiculata Lindl.
Evening Primrose Family. ONAGRACEAE

The Elegant Clarkia is an annual from 1 to 6 feet high. The erect stems and alternate, toothed, lance-shaped, inch-long leaves are smooth and gray-green. The brilliant, purple-pink petals have a distinct diamond-shape above and are narrowed below to a stalk-like claw. The capsules are covered with long, spreading hairs. The 8 stamens have 4 red anthers and 4 creamy anthers.

C. unguiculata is frequent and often quite showy when it occurs in masses as it does on open slopes away from the coast. It blooms from April to June.

The orange-scarlet touch of color in the anthers in combination wtih the purple-pink petals and deeper purple color of the sepals provide a striking combination, unusual and memorable.

Unguiculata is Latin for "little red nail or claw" and refers to the unusual claw on the petals.

Northern Willowherb
Epilobium adenocaulon Hausskn.
Evening Primrose Family. ONAGRACEAE

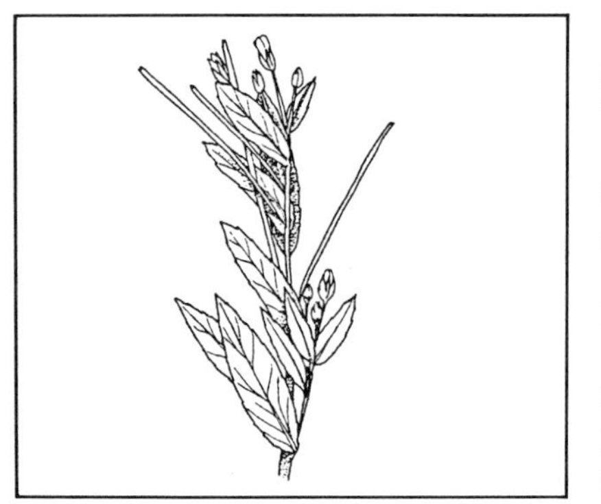

The Northern Willowherb is a perennial up to 3½ feet high with soft, appressed hairs on the stems. The toothed, oval, whitened leaves are 1 to 2½ inches long. The flowers, less than ¼ inch long, are pale pink to purplish with four 2-cleft petals. The reddish fruit pods are 1½ to 2½ inches in length. The seeds have a tuft of hair at one end.

E. adenocaulon blooms from June to October in scattered wet places at low elevations throughout.

There are many showy species of Willowherbs in the state, but not in the Santa Monicas. Fireweed *(E. angustifolium),* with its great spikes of bright pink flowers, is particularly abundant elsewhere. Fireweed appears quickly on recent burns and is said to have covered the bombed areas of London almost before the ashes were cool.

Epilobium is from two Greek words meaning "upon" and "a capsule," as the flower and capsule appear together. *Adenocaulon* refers to the small depressions on the stem.

Yellow Water Weed
Ludwigia peploides (H.B.K.) Raven.
Evening Primrose Family. ONAGRACEAE

Stanley J. Higgins

The Yellow Water Weed is a perennial herb that either floats in water or grows prostrate on muddy shores. The shiny, lance-shaped leaves are 1 to 2½ long on little ½ inch stems which may be even longer on the floating leaves. The 5, glossy, oval, yellow petals, almost an inch long, show pinnate veination. There are 8 to 10 stamens in two series. The capsule, twice as long as the petals, has conspicuous seeds buried in fleshy walls.

L. peploides is beautiful and abundant in lakes and marshy places away from the coast. It blooms from May to October.

Although there are three other species of Ludwigia in Southern California, Yellow Water Weed is the only one to be found in the Santa Monicas. A species in the eastern part of the U.S. is called False Loosestrife and another is known as Rattle Box.

Christian Gottlieb Ludwig (1709-1773) was a professor in Leipzig. *Peploides* means "resembling *Euphorbia peplus*" and describes the appearance of the plant when it grows on exposed mud, rather than under water.

Hooker's Evening Primrose

Oenothera hookeri T. & G. ssp. *grisea* (Bartlett) Munz.
Evening Primrose Family. ONAGRACEAE

Dede Gilman

Hooker's Evening Primrose is an erect perennial herb 1½ to 4 feet high with a reddish stem covered with soft, short hairs. The lance-shaped leaves are 4 to 9 inches long. There are large, showy, creamy-yellow flowers, 1 to 1½ inches long with 4 petals and 4 sepals. The sepals bend down.

O. hookeri inhabits moist places in the mountains but is not often seen. It blooms from June to August.

All Evening Primroses open suddenly with a quick motion that some claim can not only be seen but actually heard. *O. hookeri* usually has closed petals until sundown when it opens and releases its perfume. The large, lovely blossoms and sweet fragrance are designed to attract moth pollinators. It is a vigorous plant, easily grown from broadcast seed and will spread out and naturalize in sun with ample moisture.

Oenothera is from Greek, *oinus,* meaning "wine" and *thera,* meaning "to imbibe" because an allied European plant was thought to induce a taste for wine. Sir William Jackson Hooker (1785-1865) was the director of the Royal Botanic Gardens at Kew in the mid-19th century. He expanded the gardens, sent numerous botanists to many parts of the world and wrote several botanical works.

California Fuchsia

Zauschneria californica Presl.
Evening Primrose Family. ONAGRACEAE

James P. Kenney

The California Fuchsia is a slightly woody, perennial herb 1 to 3 feet high. The gray-green, very narrow leaves alternate on the main stem. The flowers grow in loose clusters on small stems on each side of the main stem. The 4 petals, 1 to 1½ inches long, are shaped like a funnel. Both the petals and the 4-cleft sepal cup are brilliant scarlet. The 8 stamens and the style hang far out. The capsule contains many seeds.

California Fuchsia was called *Balsamea* by the early Spanish who applied a solution of it to cuts and bruises. It is cultivated in gardens in the east where it is known as Hummingbird's Trumpet. It also does very well in California gardens and is readily available at some nurseries. *Z. cana* is very similar but the leaves are grayer, hairier, and narrower.

Both species are found away from the coast in dry sections of Coastal Sage, Chaparral, and Oak Woodland blooming from July through November.

The generic name honors a German botanist, Johann Baptista Josef Zauschner (1737-1799). Zauschner was a professor of medicine and botany at Prague. *Cana* means "ash-colored."

However, in recent research *Z. californica* has been placed in the genus *Epilobium*.

Oxalis Family. OXALIDACEAE

Oxalis Family consists mainly of herbs with a sour sap. The principal genus is *Oxalis* in which there are around 175 species.

The flowers are in parts of fives and tens: 5 sepals, 5 petals, 10 or 15 stamens, an ovary with 5 chambers and often 5 styles. The leaves are usually compound and alternate, or basal and opposite. The fruit is a capsule or berry-like.

Bermuda Buttercup
Oxalis pes-caprae L.
Oxalis Family. OXALIDACEAE

Geoffrey Burleigh

The Bermuda Buttercup is a smooth perennial with scaly bulbs and a deep rootstock. The 3 heart-shaped leaflets, about an inch long, are basal. The stems are 4 to 8 inches long. The showy, bright yellow flowers have 5 petals ½ to 1 inch in length. There are 5 sepals, 5 styles and 10 stamens, 5 long and 5 short. The fruit is a capsule about ½ inch long. This species does not fruit in California.

Bermuda Buttercup is a native of South Africa and is widely introduced in many places. *O. corniculata* is the troublesome garden weed with small yellow flowers. *O. rubra,* a garden escape, has rose-purple flowers. *O. albicans,* the only native member of the genus in the Santa Monicas, has densely hairy stems and is found only occasionally under brush or on rocky slopes.

Bermuda Buttercup blooms from November to April in open grassy places.

All *Oxalis* have leaves and stems with a pleasantly sour taste and can be enjoyed in salads. Take care not to consume a large quantity since these plants contain oxalic acid which may have a toxic effect.

Oxalis means "sour" in Greek. *Pes-caprae* means "foot of the goat," an allusion to the shape of the leaflet.

Peony Family. PAEONIACEAE

The Peony Family members are herbs with alternate leaves that are twice divided into three's. Petals are usually 5, sepals 5 and stamens, numerous. The fruit is heavy and fleshy and opens on one side only. It contains several large seeds.

Our only species is pictured here.

Wild Peony

Paeonia californica Nutt. ex T. & G.
Peony Family. PAEONIACEAE

Suzanne Swedo

The Wild Peony is a fleshy perennial herb 1 to 2 feet high. The pinnately divided leaves are alternate and divided at least once again. The flowers are large, solitary and terminal. There are 5 to 6 dark, blackish-red petals. The sepals are also 5 or 6 and there are numerous stamens and 2 to 5 pistils. The flowers never open wide. (Lift the nodding stems and look inside the blossom to see how interesting the Wild Peony is.) The fruit is a large fleshy follicle.

Wild Peony can be found in bloom as early as January on brushy slopes throughout. It is never abundant and its flowers are gone by April. The blossoms are erect at first but as the follicles mature, the stems begin to droop until the fruit rests on the ground.

The Spanish Californians ate the thick root raw as a remedy for stomach complaints. Indians used it powdered for colds and sore throat.

The generic name is Greek and comes from *Paeon,* the physician of the gods.

Poppy Family. PAPAVERACEAE

The Poppies are herbs, or rarely shrubs, with milky juice. The 2 or 3 sepals fall as the petals open. There are always twice as many petals as there are sepals. The numerous stamens defy counting. The fruit is a capsule with the seed attached in rows on the walls.

There are eight genera of this family in our mountains. Seven of them are described here.

Tree Poppy

Dendromecon rigida Benth.
Poppy Family. PAPAVERACEAE

Dede Gilman

Tree Poppy is 3 to 10 feet tall. The leathery leaves are 1 to 3 inches long. The bright yellow flowers have 4 petals, each an inch long. The 2 sepals fall early. There are many stamens.

D. rigida is common on dry slopes and stony washes throughout the eastern $^2/_3$ of the range. It is especially noticeable in years after a fire. It blooms from February to April.

The Tree Poppy is the only truly shrubby member of the Poppy Family here. This handsome shrub with its rigid, pale green leaves looking very much like those of the willows and its attractive bright blooms does very well in Southern California gardens.

Dendromecon is from two Greek words meaning "tree poppy." *Rigida* refers to the stiff leaves.

Ear-drops, Fire Hearts

Dicentra ochroleuca Engelm.
Poppy Family. PAPAVERACEAE

Geoffrey Burleigh

Ear-drops is an erect perennial with several thick stems up to 6 feet high. The 3-pinnately divided leaves are further dissected into narrow lobes and are scattered along the stems. The inch-long, creamy flowers with purple-tinged tips are in crowded clusters. These irregular, flattened, heart-shaped flowers have 4 petals in 2 pairs. The outer are larger with spreading tips and a sac-like base, the inner spoon-shaped and joined at the top to form a protecting dome over the 6 anthers.

Our species blooms from May to July from Sepulveda Canyon westward in Chaparral. Although you may not see this plant preceding a fire, it covers acres of the mountains after one.

A close relative, Golden Ear-drops, *(D. chrysantha),* differs in having bright yellow flowers and is common elsewhere. Both of them were once listed as members of the FUMARIACEAE, the Bleeding Heart Family.

Dicentra is from Greek and means "twice spurred." *Ochroleuca* means "yellowish-white" and is the color of the flowers.

California Poppy
Eschscholzia californica Cham.
Poppy Family. PAPAVERACEAE

Dede Gilman

The Califonia Poppy is an herb with stems up to 2 feet high. The leaves are dissected into narrow segments. The 4 satiny, showy, orange petals are up to 2½ inches long. There are numerous stamens and 1 pistil. The 2 united pistils fall early exposing a prominent, horizontal rim on the receptacle.

Although you must go to the desert areas to see the sheets of brilliance that this poppy can display, it brightens grassy slopes and flats throughout the range below 2000 feet from February to September. *E. caespitosa,* a smaller paler species that lacks the horizontal rim, blooms in April and May.

The Spanish Californians called it *Dormidera,* "the drowsy one," because the petals fold at evening. They made a hair dressing that they thought made the hair grow and shine by frying the blossoms in olive oil and adding perfume.

At maturity, the seed vessel opens abruptly with an almost audible pop and throws the seed some distance.

Dr. Johann Friedrich Eschscholtz (1793-1831) was a surgeon and naturalist who came with the Russian expeditions to the Pacific Coast in 1816 and 1824. The type specimen of this species is in an herbarium in Leningrad.

Small-flowered Meconella
Meconella denticulata Greene.
Poppy Family. PAPAVERACEAE

Linda Hardie-Scott

The Small-flowered Meconella is a smooth annual 2 to 12 inches high with oval, toothed, basal leaves 1 to 1½ inches long. The leaves higher on the stem are smaller and almost linear. The 6-petaled, white flowers are only $^3/_{16}$ inch long. The 3 sepals are slightly purplish. The fruit is a narrow, twisted capsule from ¾ to 1¼ inches long.

The Small-flowered Meconella is an inconspicuous little plant that escapes notice even when present.

It blooms in March and April on moist slopes in Chaparral.

The generic name is from Greek, *mekon* meaning "poppy" and *ella,* "a diminutive." *Denticulata* means "fine-toothed."

Fire Poppy
Papaver californicum Gray.
Poppy Family. PAPAVERACEAE

Geoffrey Burleigh

The Fire Poppy is an erect, usually smooth annual with milky sap and is about a foot tall. The leaves are pinnately dissected with lobed segments. The 4 petals, about ¾ inch long, are brick red with a greenish spot at the base. The fruit is a capsule shaped like a top.

P. californicum occupies Oak Woodland and Chaparral in the central portion of the range and blooms in April and May.

The lovely Fire Poppy is aptly named because in the spring following a fire it is quite abundant, otherwise it is seldom found. It closely resembles another almost red member of the PAPAVERACEAE, the Wind Poppy *(Stylomecon heterophylla).* The Wind Poppy, reported only from the vicinity of Malibu Lake, has yellow juice and a purple spot above a greenish base on each petal.

Papaver is the classical name of the poppy. The Latin comes from the same root as the old Celtic word *pap,* possibly because babies were given food containing poppy juice to make them sleep. There is another interpretation which suggests that the ancient Romans derived the name from the sound they made when chewing the seeds.

Cream Cups
Platystemon californicus Benth.
Poppy Family. PAPAVERACEAE

Dede Gilman

Cream Cups are low annuals with many stems from the base. The hairy leaves, mainly on the lower part of the plant, are ¾ to 2 inches long. The creamy flowers are solitary on long stems. The 6 petals are ¼ to ⅝ inch long. There are numerous stamens and 1 pistil. The 3 sepals have long hairs that make the buds hairy.

P. californicus is not frequently encountered. It occurs in grassy areas away from the coast blooming in March, April and May. Look for it in Griffith Park, around Lake Sherwood, in Zuma Canyon, and Malibu Creek State Park.

Some people like the odor of Cream Cups while others find it unpleasant. The subtle tones of pale straw to cream-colored yellow seen in this flower make it attractive without debate, however.

The genus, *Platystemon,* takes its name from the flattened stalks of the stamens.

Matilija Poppy
Romneya coulteri Harv.
Poppy Family. PAPAVERACEAE

Linda Hardie-Scott

The Matilija Poppy is a perennial, 3 to 7 feet tall, woody at the base. The alternate leaves are gray-green and pinnately divided. The large, showy flowers are at the ends of the branches. The 6 white, crinkled petals are 1½ to 4 inches long. There are 3 sepals and many stamens creating a large yellow center.

There is a large population in Malibu Creek State Park and another reported at Saddle Peak. It blooms from May to July.

The Matilija Poppy is native in Southern California and in Baja California, but has been introduced to the Santa Monica Mountains. It is cultivated in gardens where, once established, it thrives and multiplies. It has a heady, delicious perfume and a large stand can be detected by nose even before sighting.

Dr. Romney Robinson (1792-1882) was an Irish astronomer. Dr. Thomas Coulter (1793-1843) was the botanist who first collected this outstanding plant and was in California from 1831-1832. He, like Robinson, was Irish and was also a geologist.

In dedicating this flower, William Henry Harvey (1811-1866) wrote in 1845:

> "In the collection brought by the late lamented Dr. Coulter from California, I was immediately struck by the singularity of a fine Papaveraceous plant, which I soon ascertained to be distinct from any hitherto recorded from that country; and a closer examination and conference with Sir W. J. Hooker, proved it to belong to a new and curious genus, closely allied indeed to *Papaver,* but differing in its calyx, the form of its capsule, and the disposition of its stigmata. Had there been no genus *Coulteria,* it is to this that I should have affixed the name of Dr. Coulter; but DeCandolle having in this matter long anticipated me, I desire, as the next greatest respect that I can pay to Dr. Coulter's memory, to bestow upon this fine plant of his discovery, the name of his most distinguished and one of his most intimate friends. I therefore propose to inscribe it to the Rev. Dr. T. Romney Robinson, the Astronomer of Armagh; not that I have the vanity to suppose that my doing so can add any ray to the name of Romney Robinson, a name already caught up in the stars; but simply to indulge the wish, above expressed, of honoring Dr. Coulter's memory through his friend, and thus linking the names of Coulter and Romney Robinson as closely in the annals of science as their friendship was strong and indissoluble. I regret that an elder *Robinsonia* prevents me from making use of Dr. Robinson's family name; but in calling this present genus *Romneya,* I follow a sufficiently established precedent."

Plantain Family. PLANTAGINACEAE

There are but three genera in this family and the largest one is *Plantago.* There are 14 species of this genus in California, 6 of which are introduced. All of these are low herbs with basal leaves and tightly clustered flowers on long stems.

Common Plantain

Plantago major L.
Plantain Family. PLANTAGINACEAE

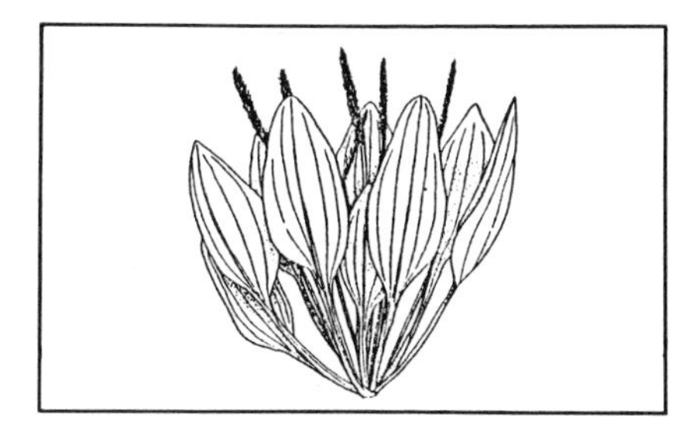

The Common Plantain is a perennial plant with thick, oval, basal leaves up to 6 inches long and conspicuously ribbed from base to apex. The leafless, flowering spikes, 2 to 16 inches long, are erect with small, 4-petaled blossoms that are followed by many, small brown capsules.

Common Plantain, sometimes known as Goosetongue, is a native of Europe and Asia and is now distributed worldwide. It may be found in moist places throughout, blooming from May to November.

P. lanceolata, another immigrant, known as English Plantain or Ribwort, has a dense flowering spike that is cone-shaped at the apex. *P. erecta,* our native species which prefers a dry habitat, has shorter, more rounded, flowering spikes and linear leaves.

The seeds of *P. major* have been found in the tombs of Pharaohs stored for food. The American Indians called it "White Man's Foot," since it appeared shortly after every new incursion of Europeans. Young leaves are tasty, boiled, or eaten fresh in salads.

Plantago is from Latin meaning "footprint" which is a curious coincidence with the American Indian name. *Major* means "larger" and refers to the leaves.

Leadwort Family. PLUMBAGINACEAE

The Leadwort Family members are perennial herbs, shrubs and woody vines. The leaves are simple, often in basal rosettes. The 5-parted flowers are often in heads on long stalks. The petals are somewhat joined at the base and are folded over in the bud. The stamens are opposite the center of the petals. The pistil has five styles. The ovary is superior and there is a single seed.

As you might imagine, the cultivated shrub, Plumbago, that thrives in Southern California gardens, belongs to this family as does Statice, which like our native species seems to last forever in a dry condition without losing shape or color.

Sea Lavender

Limonium californicum (Boiss.) Heller.
Leadwort Family. PLUMBAGINACEAE

Sea Lavender is a perennial herb 8 to 20 inches tall with large, basal leaves up to 8 inches long on stems almost as long. The branches are densely flowered at the tip. The petals are pale violet.

Our native species may be found in salt marshes along the coast, notably at Point Mugu blooming from July until December.

There are 3 species of *Limonium* in the state but this is the only one which is a true native. The others have been introduced from the Old World. *L. perezii* has pale yellow petals and has naturalized along our coast from its home in the Canary Islands.

Limonium comes from an ancient Greek name for "a marsh."

Phlox Family. POLEMONIACEAE

The Phlox are herbs and shrubs with 5 narrow sepals and 5-lobed petals which form a tube or a funnel. The 5 stamens are on the flower tube. The style is 3-lobed. There is a superior ovary and the fruit is a small capsule.

Many members of this family are well-known garden flowers and wildflowers in North America. There are more native species in California than elsewhere.

There are seven genera in the Santa Monicas.

Sapphire Eriastrum

Eriastrum sapphirinum (Eastw.) Mason.
Phlox Family. POLEMONIACEAE

Geoffrey Burleigh

The Sapphire Eriastrum is a sticky, erect annual, 4 to 12 inches high with alternate, linear leaves up to an inch long. The deep blue flowers, funnel-shaped with a yellow tube and throat, are ½ inch long. The fruit is a small, several-seeded capsule.

E. sapphirinum is scattered in Chaparral and on rocky slopes away from the ocean throughout. It blooms in May and June.

The other two species of *Eriastrum* reported from the Santa Monicas, *E. densifolium* and *E. filifolium,* are almost never encountered. Both prefer much higher elevations.

Eriastrum is from two Greek words meaning the plants are "woolly with star-like flowers." *Sapphirinum* means "blue."

Angeles Gilia

Gilia angelensis V. Grant.
Phlox Family. POLEMONIACEAE

The Angeles Gilia is a low, slender annual 3 to 25 inches high. The alternate leaves are dissected and up to 2 inches long. The pale violet flowers, bell-shaped with a yellow tube, are almost ½ inch long. The small oval capsules produce 20-30 seeds.

G. angelensis is frequent in Grassland and open areas in Chaparral away from the coast. It blooms in April and May.

There are many lovely wildflowers in the genus *Gilia.* The brilliant, showy species of the deserts and of higher elevations are not well represented here. *G. angelensis* is pale and slight, and *G. capitata* with powder blue flowers in a head occurs in masses only after fires.

The genus was named for Filippo Luigi Gilli (1756-1821) who collaborated on a botanical work with the Spaniard Caspar Xuarez (born in Paraguay in 1731 and died in Rome in 1804). Xuarez was both a botanist and a Jesuit. *Angelensis* is "of Los Angeles,"—Los Angeles County, not the city.

Blue-headed Gilia, Globe Gilia

Gilia capitata Sims ssp. *abrontanifolia* (Nutt. ex Greene) V. Grant.
Phlox Family. POLEMONIACEAE

Suzanne Swedo

The Blue-headed Gilia is a tall, slender annual 8 to 32 inches high. The finely dissected, basal leaves are up to 4 inches long; the upper ones are much reduced. The small, blue flowers with exserted stamens are crowded in head-like clusters on long naked stalks. The fruit is an oval capsule.

Blue-headed Gilia is occasionally found in open places, on banks, in burned and disturbed areas in Chaparral and Oak Woodland at low elevations. It blooms in April and May.

G. capitata often shows up in commercial wildflower mixtures. The seeds germinate most freely in early fall rains. The plants may come up again next year if allowed to go to seed. They prefer sun, lean soil and can tolerate small amounts of water.

Capitata refers to the rounded head of clustered flowers.

Prickly Phlox
Leptodactylon californicum H. & A.
Phlox Family. POLEMONIACEAE

Linda Hardie-Scott

The Prickly Phlox is a much-branched shrub about 3 feet tall with woolly stems. The leaves which are rigid, sharp-pointed and palmately lobed fall off leaving smaller, prickly, needle-like leaves that grow in bunches. The rose-pink flowers, 1 to 1½ inches across, have a white center and a long tube that expands into a flat border. They are solitary or in clusters of a few blossoms at the end of the branchlets.

Prickly Phlox is common in openings in Chaparral and blooms from February to May.

The satin sheen of the bright blossoms is beautifully set off by the dark, prickly armor. There is a delicate fragrance. One common name is Mountain Pink, since they like to cling on dry hillsides. Another is Rock Rose. Both of these are unfortunate since both of these names also belong to very different families.

Leptodactylon in Greek means "narrow fingers" and refers to the leaves.

Ground Pink
Linanthus dianthiflorus (Benth.) Greene ssp. *dianthiflorus.*
Phlox Family. POLEMONIACEAE

Margaret Stassforth

The Ground Pink is a very low annual with slender stems, 2 to 4½ inches high, with very fine, smooth-margined, opposite leaves. The large, showy, mostly solitary flowers, up to an inch long, are lilac or purplish with a dark-spotted, yellow throat. The petals are fringe-toothed.

The Ground Pink is not common but can be found as a colorful delight on grassy slopes in open Coastal Sage at low elevations. It blooms from February to April.

This is one of the choicest of spring annuals. The vital energy of the little plant goes into the generous flowers which quite hide the leaves and stems. It has been introduced into cultivation both in this country and abroad under the name *Fenzlia dianthiflora* and several varieties are used in edgings and in rock gardens.

The generic name comes from Greek and means "flax flower." *Dianthiflorus* means that these flowers resemble carnations.

Hooked Navarretia

Navarretia hamata Greene.
Phlox Family. POLEMONIACEAE

Linda Hardie-Scott

The Hooked Navarretia is an erect annual up to 5 inches tall. The clasping leaves have 2 or 3 lobes on each side. The funnelform, purple flowers are in small heads subtended by broad bracts. The exserted stamens are unequal in length. The fruit is a small capsule.

Hooked Navarretia is scattered in Chaparral back of the immediate coast throughout. It blooms in May and June.

The downward lobes of the leaves seem to form hooks giving rise to both the common name and the scientific one. This is a western American genus, 29 of its 30 species being found in western North America. The only other *Navarretia* in our mountains is *N. pubescens,* a sticky-hairy species which has been seen recently in Cheeseboro Canyon.

Francisco Fernandez de Navarrete (d. 1742) was an 18th century philosopher, anatomist, naturalist and physician to Felipe V of Spain. *Hamata* means "hooked."

Buckwheat Family. POLYGONACEAE

Plants in the Buckwheat Family are herbs or woody plants with swollen joints. A conspicuous characteristic of the Buckwheat Family is the sheath that encircles the stem at every point where a leaf is attached, but this does not occur in *Eriogonum* or *Chorizanthe.* The flowers have no petals but there are 3 to 6 sepals which resemble petals. There are 3 to 9 stamens, a superior ovary and the dry fruit is one-seeded.

Rhubarb for pies and buckwheat for flour belong in this family.

Turkish Rugging

Chorizanthe staticoides Benth.
Buckwheat Family. POLYGONACEAE

Suzanne Swedo

Turkish Rugging is a low annual with brittle, much-branched, rosy stems. The small, reddish or white flowers, 1 to 3 of them in a cluster, are enclosed in a cup-like bract with hooked spines. The leaves are in a basal rosette and disappear early.

Look for Turkish Rugging on grassy slopes and openings in brush throughout. It is in bloom from April to June.

This plant is noticed most in the dry days of summer on arid hills and plains where spreading low over considerable areas it lends its rosy color to the landscape.

Chorizanthe, from Greek, means "divided flowers," but refers to the divided calyx. *Staticoides* means "like *Statice." Statice* is a synonym of *Limonium* which is a genus of PLUMBAGINACEAE.

California Buckwheat.

Eriogonum fasciculatum Benth.
Buckwheat Family. POLYGONACEAE

Dede Gilman

The California Buckwheat is a small shrub 2 or 3 feet high with little, pinkish flowers in dense terminal heads. The many, narrow leaves are in bundles all along the stems.

California Buckwheat may be found in Coastal Sage and Chaparral throughout and blooms from April to October.

California Buckwheat is one of the most characteristic plants of the Southern California Chaparral. It is cherished by bees and makes an exceptionally fine honey. The

Suzanne Swedo

warm-brown of the fruiting heads gives color to the drying landscape. Even though the buckwheat of flour fame belongs in the same family, an attempt to make anything palatable of this plant is probably doomed to failure.

Eriogonum is from two Greek words meaning "woolly plant" and "bent abruptly as a knee." The plants may be hairy at the nodes. *Fasciculatum* is a Latin word meaning "bundles" and describes the growth habit of the leaves.

In North America there are perhaps 150 species of *Eriogonum*. Most are in the West. There are eight recorded in the Santa Monica Mountains. The ones most likely to be encountered are listed here.

Ashy Leaf Buckwheat *(E. cinereum)* has oval leaves and a gray appearance. It is most abundant near the coast.

Long-stemmed Eriogonum *(E. elongatum)* has clumps of erect, bare, flowering stems. The leaves wither by flowering time. It is common throughout.

Conejo Buckwheat *(E. crocatum)* was listed several years ago as rare and endangered. It has since been thriving and new sitings in the Ventura County areas of the mountains have been recorded. It has striking yellow puff balls of bloom from April to July.

Water Smartweed
Polygonum punctatum Ell.
Buckwheat Family. POLYGONACEAE

The Water Smartweed is an erect perennial 1 to 3½ feet high. The lance-shaped leaves are 2 to 4 inches long. The small, greenish-white flowers are borne in small clusters on a long spike with the younger ones at the top.

Water Smartweed is common in marshy ground at the western end of the mountains. It blooms from July to October.

The genus *Polygonum* includes plants known as Knotweeds, Black Bindweed and many other Smartweeds. Six *Polygonums* are reported in the Santa Monicas. *P. arenastrum* from Eurasia has a prostrate, sprawling habit and does especially well in hard ground. *P. coccineum* can be seen in dense clumps around Lake Sherwood. This one is a great pest in the Sacramento Valley delta region where it is called "kelp." The other three species are rarely encountered in our mountains.

Polygonum is derived from a pair of Greek words meaning "many joints" because of the thickened joints on the stem. *Punctatum* refers to the gland-dotted sepals.

Curly Dock
Rumex crispus L.
Buckwheat Family. POLYGONACEAE

Barry Silver

Curly Dock is a perennial with a stout stem from 1½ to 4 feet high. In its first year it forms a dense rosette of bluish-green leaves around a stout taproot. The curly, wavy leaves are 3 to 10 inches long on little stalks 1 to 2 inches long. The dense inflorescence is 1 to 2 feet long, leafless, and rose-colored when fruiting. The inner sepals are heart-shaped.

Curly Dock is common in waste places and along roads throughout. It blooms from March to June. It was introduced from Europe and Asia and is widely distributed throughout the United States.

There are 5 species of *Rumex* in the Santa Monicas but only two are likely to be encountered. Willow Dock *(R. salicifolius)* has a leaf surface that is flat rather than crinkly as in Curly Dock, and *R. conglomeratus* resembles a more slender *R. crispus*.

Rumex is the ancient Latin name for the docks or sorrels. *Crispus* is also from Latin meaning "curled."

Purslane Family. PORTULACACEAE

Plants in the Purslane Family are small succulent herbs with simple, smooth-margined leaves. There are 2 sepals, 5 to many petals and few to many stamens. The fruit is a capsule.

Miner's Lettuce and Red Maids are the most common of our native species.

Red Maids

Calandrinia ciliata (R. & P.) DC. var. *menziesii* (Hook.) Macbr.
Purslane Family. PORTULACACEAE

Dede Gilman

Red Maids are low annuals, 14 to 16 inches high. The glossy, magenta flowers, up to ¾ inch across, have 5 petals and 2 sepals. The alternate leaves are somewhat narrow, smooth-margined and well distributed on long leaf stems.

Red Maids bloom from February to June on grassy slopes throughout. They are seldom found in abundance but are easy to locate in the grass in late spring and are easy to recognize by their little, brilliant, satiny petals.

Calandrini (1703-1758), a professor of mathematics and philosophy, was an author of botany in Geneva, Switzerland. *Ciliata* was given to indicate the slight fringing of the petals.

Miner's Lettuce

Claytonia perfoliata Donn.
Purslane Family. PORTULACACEAE

Frances Vogel

Miner's Lettuce is a low, succulent annual a few inches to a foot high. The white flowers have 5 petals, 2 sepals and 5 stamens. The leaves form a circular disk that completely surrounds the stem.

C. perfoliata is common in shaded and vernally moist places and blooms from February to May.

Miner's Lettuce is said to be palatable raw or boiled with salt and pepper. Indians in Placer County are reported to have put the leaves near the entrance of red ant holes.

After the ants had swarmed on the leaves, they were shaken off leaving a vinegary taste which was much relished.

John Clayton (1693-1773), an American botanist and the Attorney General for Colonial Virginia, was the sole collector for an early flora of Virginia. The manner in which the stem "perforates" the leaf is most distinctive and is expressed in the species name, *perfoliata*.

Miner's Lettuce was once placed in the genus *Montia*, named for Guiseppe Monti (1682-1760) who was a professor of botany at Bologna, Italy.

Primrose Family. PRIMULACEAE

Plants in the Primrose Family are herbs with single leaves and regular flowers parted in 5's. The petals are joined and the 5 stamens are opposite the center of each petal. The fruit is a capsule.

Primroses, Cyclamens and Cowslips are in this family, but none of these are found in our mountains. The weedy, little Scarlet Pimpernel *(Anagallis arvensis)*, which is often seen, is not a native. It comes from Eurasia. The Shooting Star is our only true native member of the PRIMULACEAE.

Scarlet Pimpernel

Anagallis arvensis L.
Primrose Family. PRIMULACEAE

Margaret Stassforth

The Scarlet Pimpernel is a low, spreading annual of neat habit with opposite, oval leaves growing directly from square stems. The dark, salmon-colored, wheel-shaped flowers, barely ½ inch in diameter, have a dark purple spot at the center and rise singly on thread-like footstalks from the leaf axils.

This immigrant from the Old World has become quite at home all across the United

States. Scarlet Pimpernel blooms from February through October in disturbed places.

The flowers close at nightfall or even in cloudy weather. For this reason it has a number of descriptively amusing common names in England, such as Poor Man's Weather Glass, Wink-a-peep and John-go-to-bed-at-noon. The plant contains an acrid poison and was used medicinally as a remedy for plague, gout, convulsions and hydrophobia in ancient times.

Anagallis is composed of two Greek words meaning "again" and "to delight in" since the flowers open each time the sun strikes them. *Arvensis* means "of the fields."

Padres' Shooting Star
Dodecatheon clevelandii Greene.
Purslane Family. PORTULACACEAE

Linda Hardie-Scott

The Padres' Shooting Star is a slightly-sticky perennial covered with soft, short hairs. The naked stalk grows from a few inches to a foot or more tall. The thickish leaves appear in a tuft at the base. The flowers hang down from stalks which are attached close together at the top of the stem. The pink petals with a yellow ring at the base are sharply bent back so that they point upward like the hothouse Cyclamen, and like the Cyclamen, they have a delicate scent. The stamens and pistil are bunched together and extend like a purple beak at the tip of the nodding flowers.

Once seen, the Shooting Star is unique enough to be remembered always. It is exquisite. A game called "Rooster Heads" was once played by early Spanish American youngsters—they hooked two blooms together to see which would pull away. It is not recommended today. Those plants that still survive should be admired, not destroyed.

The Shooting Star blooms in grassy, damp areas from February to April.

Dodecatheon was the name of a plant whose description suggests the Primrose. The literal meaning, "twelve gods," implied that it was a powerful medicine. Daniel Cleveland (1838-1929) was an attorney, founder of the Bank of San Diego, founder of the San Diego Natural History Society, an authority on ferns and a botanical collector. He once made a special project of rediscovering all the plants in the San Diego area that had been located or found only once.

Buttercup Family. RANUNCULACEAE

Plants in the Buttercup Family are usually herbs, although *Clematis* is a woody vine. The petals on flowers of this family are of an indeterminate number or lacking, the sepals have a petal-like appearance. The stamens are likewise indeterminate in number, but generally are many. The leaves are commonly divided or cut.

Of our four genera, only the Meadow Rue *(Thalictrum polycarpum)* is not graced with showy blossoms.

Clematis, Virgin's Bower

Clematis ligusticifolia Nutt. in T. & G.
Buttercup Family. RANUNCULACEAE

Geoffrey Burleigh

Virgin's Bower is a woody, twining vine with opposite, compound leaves of 5 to 7 leaflets which are oval or lance-shaped with 3 coarsely-toothed lobes. There are no petals; the 4 or 5 rich, creamy sepals only resemble petals. The stamens are numerous. The many pistils develop long silky tails that twist together into soft balls which are more noticeable than the blossoms.

C. ligusticifolia climbs over trees in canyons, often near streams, from Topanga Canyon eastward and blooms from April to June.

C. lasiantha is also common in chaparral canyons but its leaves have only 3 to 5 leaflets. Its flowers, on long flower stems, are fewer but more showy. The petal-like sepals of this species are hairy. It blooms from February to May.

The early Spanish valued this plant, calling it *Yerba de Chibato,* "herb of the young kid." Goatherds and shepherds made a wash of it to cleanse and cure the wounds of their animals.

The generic name in Greek means "long, lithe branches" and the species name is from Latin meaning "leaves like those of Lovage *(Ligusticum).*"

Scarlet Larkspur

Delphinium cardinale Hook.
Buttercup Family. RANUNCULACEAE

Janet Vail

The Scarlet Larkspur is a robust perennial with erect stems 2 to 6 feet tall. As in all Larkspurs, the upper sepal elongates into a noticeable spur. The petals and sepals are blazing red. The leaves, mostly withered at the time of flowering, are palmately 5-parted; the divisions are then further divided into narrow parts.

Scarlet Larkspur is common on brushy slopes and roadbanks away from the immediate coast and blooms in June and July.

Some Indians are reported to have used the root of this plant to put their gambling opponents to sleep!

Delphinium is derived from the Latin for "dolphin" because of the flower shape in some species. *Cardinale* is "red," of course.

Blue Larkspur

Delphinium parryi Gray.
Buttercup Family. RANUNCULACEAE

Geoffrey Burleigh

The Blue Larkspur is a slender perennial with a simple stem 1 to 3 feet tall. The palmate leaves are few and further divided. The blossoms are on flower stems up to 1½ inches long. The sepals are purple-blue and the straight spur is the same length as the sepals. The upper petals are lighter blue or whitish. There are many stamens.

D. parryi is frequent in Grassland throughout away from the coast. It blooms in June and July. It is gratifying to discover its rich blue color in the tall, yellowing grass.

We also have *D. patens*, which is very similar, but prefers shaded banks in oak woodlands and blooms earlier, from March to May.

Some Larkspurs are poisonous to sheep and cattle. The early Spanish called them *Espuela del caballero*, "Cavalier's Spur."

The species name honors Dr. Charles Christopher Parry (1823-1890) who was a botanist with the Mexican Boundary Survey and the Pacific Railway Survey. Besides working with the Surveys, he made independent visits to the mountains and deserts of the Southwest. He collected in the mountains east of San Diego, the deserts of Arizona, New Mexico and California and the Rocky Mountains. He is commemorated in more than a dozen California plant names.

California Buttercup

Ranunculus californicus Benth. var. *californicus*.
Buttercup Family. RANUNCULACEAE

Frances Vogel

The California Buttercup is an erect perennial, 1 to 2 feet high. The leaves are divided into 3's with the lobes then divided again. The flowers are shiny bright yellow with 9 to 16 petals and green sepals turned backwards. There are many stamens. The high gloss of the petals is the most distinctive feature. They look as if they have been waxed.

California Buttercup is found occasionally on grassy slopes below 2000 feet throughout and blooms from February to April.

Indians parched the seeds by tossing them in flat baskets with hot embers and pebbles. The seeds were then ground into meal, which they ate without further cooking. The taste is reported to resemble that of parched corn.

Ranunculus is from Latin and means "little frog," given because of the moist habitat of many species.

Many-fruited Meadow Rue
Thalictrum polycarpum (Torr.) Wats.
Buttercup Family. RANUNCULACEAE

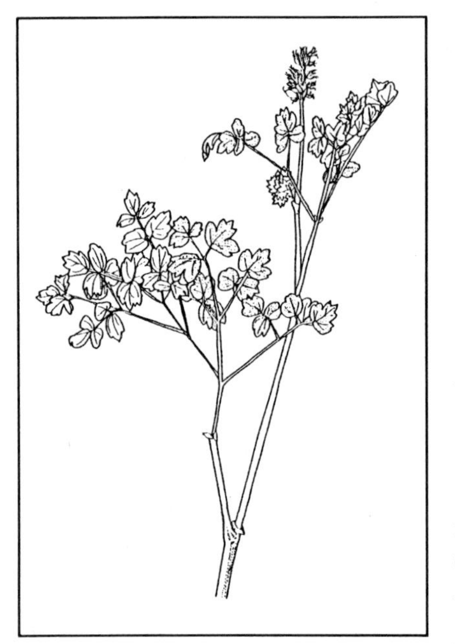

The Many-fruited Meadow Rue is a smooth, hairless, coarse perennial 3 to 8 feet tall. The leaves are twice-compound and divided into three's. There are no petals and the sepals are green. There are many drooping stamens on the male flowers and a slender purple style on the female flowers. Male flowers and female flowers occur on separate plants.

T. polycarpum is scattered in colonies along streams in cool canyons and meadows and on brushy slopes. It blooms in April and May.

Many-fruited Meadow Rue, like all the members of its genus, is a soft harmony of fresh greens, browns and purples. It is not a showy plant, but its broad airy foliage is as graceful as it is reminiscent of the Maidenhair Fern, making it a graceful addition to the home garden.

Thalictrum is the name given to the genus by Dioscorides, the Greek physician and pharmacologist whose work, *Materia Medica,* (c. A.D. 70) was the foremost classical source of modern botanical terminology and the leading pharmacological text for sixteen centuries. *Thalictrum* means "to flourish, or look green, being full of leaves." *Polycarpum* means "many seeds."

Buckthorn Family. RHAMNACEAE

Members of the Buckthorn Family are shrubs or small trees with simple leaves and small flowers. The petals, sepal lobes and stamens are 4 or 5 each. The stamens are opposite the petals.

This family is well represented in the Santa Monica Mountains with three species of *Rhamnus* and six of *Ceanothus.* Many of the plants begin to bloom early in the spring. The genus *Rhamnus* has soft, berry-like fruit while the fruit of *Ceanothus* is a hard, nut-like capsule.

Greenbark, Redheart

Ceanothus spinosus Nutt. in T. & G.
Buckthorn Family. RHAMNACEAE

Tim Thomas

Greenbark is a shrub 6 to 18 feet tall. The smooth, olive-green bark gives rise to one of its common names and is a fairly sure way to distinguish this species from the 5 others that grow in the Santa Monica Mountains. The ends of the branches are rigid and spiny. The oval leaves, no more than an inch wide, have smooth margins, no hairs and one main vein from the base. The individual, pale blue flowers are small, but showy in large 6 inch long clusters. The 3 lobes of the fruit do not have horns as do the fruits of some other species.

Greenbark is our most easily recognizable species and is common throughout. It blooms from February to May.

The name California Lilac or Wild Lilac is often applied to the score or more species in this genus that are indigenous to the Pacific Coast. Soap Bush, too, is a name often encountered since the flowers develop a bit of a cleansing lather when rubbed for a moment or two in water.

Ceanothus in Greek is "spiny plant." *Spinosus* means that this one is even more spiny.

There are six species of *Ceanothus* in the Santa Monicas. This small chart is designed to identify them.

1. The small leaf-like structures called stipules at the base of the leaf or leaf stem are thin and fall early. The leaves are smooth and the fruit is not horned.

A. Hairy-leaf Ceanothus *(C. oliganthus):* 3-veined leaf with a toothed edge and covered with soft short hairs on the top; blue flowers; flexible branchlets; blooms March and April; especially abundant on Castro Peak.

B. Greenbark *(C. spinosus):* one-veined leaf; rigid branchlets; flowers white to pale blue; blooms February to May.

C. Chaparral White Thorn *(C. leucodermis):* the top of the smooth-edged leaves is covered with a white delicate powdery coating that can be rubbed off; blooms in April and May; flowers white to blue; not seen often.

2. The stipules are thick and hang on; the leaves are rough and have tiny holes (stomata) on the underside; horns are on the fruit; the flowers are always white.

A. Big Pod Ceanothus *(C. megacarpus):* leaves alternate; often composes 50 percent of the cover of southern slopes; blooms January to April.

B. Buck Brush *(C. cuneatus):* leaves opposite; blooms February to April.

C. Hoary-leafed Ceanothus *(C. crassifolius):* leaves opposite with midrib densely woolly on lower suface; blooms in March and April.

Coffeeberry
Rhamnus californica Esch. ssp. *californica.*
Buckthorn Family. RHAMNACEAE

Linda Hardie-Scott

The Coffeeberry is a shrub 3 to 12 feet tall. The oblong leaves, 1 to 3 inches long, are shiny above and have turned-inward margins. The greenish flowers, only ⅛ inch across, have 5 petals which are shorter than the 5 sepals. The flowers are numerous and are on small stalks that originate from a central point. The green, black or red fruit is a soft berry, shaped like the commercial coffee bean.

Coffeeberry occurs in Oak Woodland and Chaparral throughout the area. It blooms in May and June.

Redberry *(R. crocea)* has 4 petals and blooms in March and April. Holly-leaf Coffeeberry *(R. ilicifolia),* also with 4 petals and sepals, was once considered a subspecies of Redberry; however, it has longer leaves and longer flower buds than *R. crocea.*

The medicinal benefits of *R. californica* were known to the Indians who are said to have used it to correct the effects of an acorn diet. The bark was formerly widely collected and exported to be made into a laxative. It was hoped that the berries would be a perfect substitute for coffee, but the flavor did not live up the shape of the berry.

The genus *Rhamnus* is often known as Buckthorn or Cascara. *Rhamnus* is the ancient Greek name for the Buckthorn.

Rose Family. ROSACEAE

This family is a hard one to single out at a quick glance. Look for a pair of small, leaf-like structures, called stipules, at the base of the alternate leaves; flower parts attached to a disk or cup, called a hypanthium; 5 separate petals and sepals; numerous stamens. Family members may be herbs, vines, shrubs or trees. The fruit type varies greatly.

Many of our favorite fruits, such as apples, pears, cherries and strawberries, belong to the Rose Family. There are eleven genera in our mountains, most of them shrubs or small trees. Chamise *(Adenostoma fasciculatum)* and Toyon *(Heteromeles arbutifolia)* are widespread, bushy shrubs typical of our Chaparral.

Chamise, Greasewood
Adenostoma fasciculatum H. & A.
Rose Family. ROSACEAE

Dede Gilman

The Chamise is a shrub 2 to 15 feet high with evergreen, needle-like, ¼ inch leaves in clusters along the branches. The white flowers, individually very small, are disposed in showy, crowded, compound clusters several inches in height and terminal on the branches. The fruit is an achene.

Chamise is one of the dominant plants found throughout Chaparral and Coastal Sage away from the immediate coast. It blooms from April to June. In the Coast Ranges in early spring, Chamise often covers miles of slopes with first a dense uniform green. Then in April and May, the slopes are diffused in snowy white blooms followed by a warm bronze from the abundant seed vessels.

The bark of Red Shanks or Ribbonwood *(A. sparsifolium)* peels off in long strips. The flowers are loose on little stems and its leaves are not in bundles. It is much less common than Chamise and is found at higher elevations. The Spanish Californians called this one *Yerba del Pasmo,* "herb of the convulsion," and thought that it was a remedy for such diverse ailments as colds, cramps, snakebite and lockjaw.

Greasewood is a name given to many southwestern plants. In this case, the dry branches are most flammable as though they do indeed contain grease. They often contribute to the spread of brush fires. Chamise is an Americanized version of a Spanish word which means brush or fire-wood. *Chamiso* comes from the Portuguese word, *Chama,* "a flame."

Adenostoma is from the Greek *aden,* "a gland," and *stoma,* "mouth," in reference to the 5 glands at the mouth of the sepals. *Fasciculatum* describes the needle-like leaves which appear in bundles.

California Mountain Mahogany

Cercocarpus betuloides Nutt. ex T. & G.
Rose Family. ROSACEAE

Dede Gilman

The California Mountain Mahogany is a shrub or small tree 3 to 20 feet high with small, wedge-shaped leaves. The tiny, white flowers in small clusters have the fragrance of Acacia. The fruit is a tubular achene which develops a distinctive, feathery, curling tail 2 or 3 inches long in early summer. Even without this identifying feather tail, *Cercocarpus spp.* can be recognized by the leaves which are smooth-edged from the base to the middle and toothed above. No other native shrub has this feature.

California Mountain Mahogany is common in Chaparral back of the immediate coast throughout. It blooms from March to May.

There are several species of this genus in California, but this is the only one in our mountains.

The extreme hardness of the wood gave rise to its common name. Since the Indians had no iron, they found it extremely useful for digging sticks, fish spears and arrow tips.

Cercocarpus is from Greek, *kerkos,* "tail" and *karpos,* "fruit." *Betuloides* means "like *Betula*" and refers to the leaves. *Betula* is the genus for the birch.

Toyon, Christmas Berry

Heteromeles arbutifolia M. Roem.
Rose Family. ROSACEAE

Geoffrey Burleigh

Toyon is a shrub or small tree from 6 to 25 feet high. The evergreen, leathery, rigid leaves are 2 to 4 inches long, oblong and saw-toothed. The small, white flowers have roundish petals and grow in loose compound clusters. The sepal cup is 5-toothed. There are 10 stamens. The large and abundant clusters of red berries are better known than the flowers. If you cut open a berry (more properly known as a pome), you will see that the center resembles the core of an apple with the seeds well protected inside.

Nancy Dale

Toyon blooms in June and July in Chaparral and Coastal Sage throughout. The berries, of course, come later. Under state law the branches of this species must NOT be collected.

The berries are sweet and spicy. Indians toasted or boiled them and ate them with great relish. The Spanish Californians used them to prepare a pleasant drink. It is thought that masses of this native shrub growing on the hills above Hollywood gave that community its name.

Heteromeles means "different apple." *Arbutifolia* refers to *Arbutus unedo,* the Spanish Madrone, or Strawberry Tree.

Sticky Cinquefoil

Potentilla glandulosa Lindl. ssp. *glandulosa.*
Rose Family. ROSACEAE

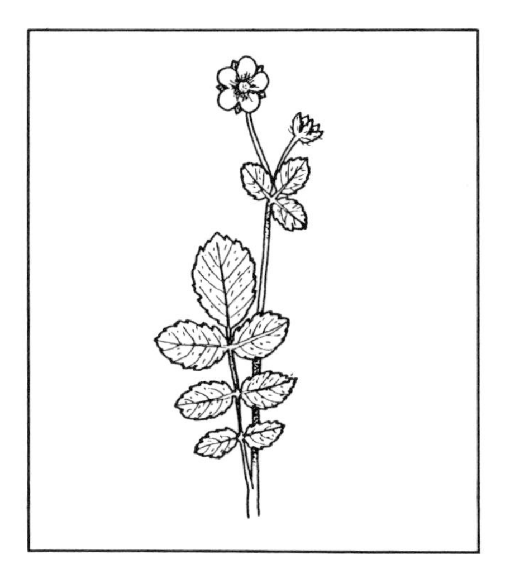

The Sticky Cinquefoil is an erect, perennial herb 1 to 3 feet tall with sticky-hairy herbage. The leaves are pinnately compound into 5 to 9 leaflets which are oval, toothed, ½ to 1½ inches long, green above and lighter beneath. The strawberry-blossom-like flowers with pale yellow to creamy-white petals are about ¼ inch long. The sepals are the same length as the petals. There are 25 stamens and many pistils. The fruit is an achene.

Sticky Cinquefoil is scattered on shaded hillsides and in canyons below 2000 feet throughout. Blossoms appear from April to June.

Many species of this genus have been described as native to California. There are only three in the Santa Monicas. Silver Weed *(P. egedei)* borders salt marshes along the coast at Santa Monica and Oxnard and has bright yellow flowers and leaves silvery beneath. *P. lindleyi* is very similar to *P. glandulosa* but with only 10 stamens.

Most of the perennial Cinquefoils will root from cuttings.

Potentilla comes from the Latin diminutive of *potens,* "powerful," in reference to the medicinal properties of some species. *Glandulosa* means "provided with glands."

Holly-leaved Cherry, Islay
Prunus ilicifolia (Nutt.) Walp.
Rose Family. ROSACEAE

James P. Kenney

The Holly-leaved Cherry is an evergreen shrub 8 to 30 feet high with thick, alternate, spiny-toothed, holly-like leaves an inch or two long. The small, white flowers appear terminally on small stalks with the youngest ones nearest the top. The purple-black fruit, a drupe, has a very thick pulp layer around a one-seeded stone.

Holly-leaved Cherry is common in Chaparral throughout and blooms from March to May.

This is the only species of *Prunus* in the Santa Monica Mountains.

The shrubs are especially beautiful in spring with their bright green new growth and profusion of feathery blooms. It is a versatile and useful plant that has been cultivated for a hundred years. It makes an excellent garden hedge and will endure twice yearly pruning. Easily grown from seed or volunteers, it likes full sun, porous soil and regular watering to get started. Bees like it and Indians fermented it into an intoxicating drink.

Prunus is the ancient Latin name for the plum. *Ilicifolia* means "holly-like leaves."

California Wild Rose
Rosa californica C. & S.
Rose Family. ROSACEAE

Geoffrey Burleigh

The California Wild Rose is a bush 3 to 10 feet high. The alternate leaves are pinnately divided into 5 to 7 oval, toothed leaflets. The prickles in stout recurved pairs are few. There are 5 green sepals and 5 pink petals. The fragrant flowers, 1 to 2 inches in diameter, may be solitary or many. There are numerous stamens. The red, berry-like fruits are open at the top and are called hips. They are said to be rich in vitamin C and can be used to make jam or tea.

California Wild Rose is scattered usually along streams throughout below 2000 feet, and blooms from April to July.

No California wildflower gave greater pleasure to the pioneering Spaniards of 200 years ago than did the Wild Rose. The diaries of the Franciscan padres contain many references to it, and, as it reminded them affectionately of the sweet roses of Spain, they called it the Rose of Castile.

Rosa is an ancient Latin name whose exact meaning has been lost in the mists of time.

California Blackberry
Rubus ursinus C. & S.
Rose Family. ROSACEAE

Linda Hardie-Scott

The California Blackberry is a vine with thick, sprawling, spiny stems often from 3 to 6 feet long. The 1 to 3 inch long, prickly leaves are so deeply lobed they may be considered as separate leaflets. The white flowers, resembling those of the rose, are about an inch across with numerous stamens. The fruiting receptacle is a conical hemisphere covered with separate pistils each of which becomes fleshy at maturity. This aggregation of many pistils forms the berry.

California Blackberry grows mostly in open places below 2000 feet throughout and blooms from February to June.

The very sweet, shiny, blackberries are difficult to gather in quantity. Deer browse on the stems and foliage and many birds nest in the thickets and enjoy the fruit—jays, pigeons, mockingbirds, sparrows, tanagers, thrashers, towhees. Indians used the dried berries alone, or mixed them with dried meat to make cakes known as pemmican.

Himalaya Berry *(R. procerus)*, a native of Europe, has five leaf divisions and is found occasionally in open weedy places. This is the preferred berry for pies.

Rubus is the Latin name for "bramble." *Ursinus* is also from Latin meaning "a bear." The connection between bears and blackberries is obvious.

Madder Family. RUBIACEAE

This is a very large family in the tropics. The species in the United States are mostly herbaceous with leaves paired or in whorls. The flower parts number from 3 to 5. The petals are joined and radially symmetric. The ovary is inferior. The fruit is various.

Quinine, coffee and gardenias are all in this family. There are three genera in California, but only one, *Galium*, in the Santa Monica Mountains, unless we count Field Madder *(Sherardia arvensis)*, an occasional invasive lawn weed from the Mediterranean found in the Los Angeles Basin.

Narrow-leaved Bedstraw
Galium angustifolium Nutt. in T. & G.
Madder Family. RUBIACEAE

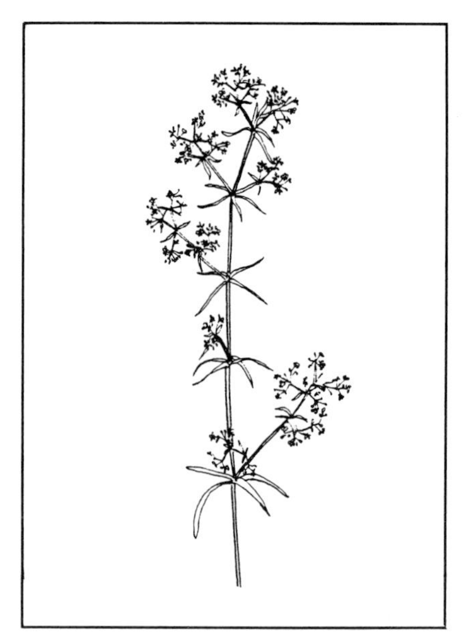

The Narrow-leaved Bedstraw is a shrubby plant 1 to 4 feet high with stiff branches and square stems. The linear, sharply-pointed leaves, ¼ to ¾ inch long, are in whorls of 4. The greenish-white flowers, ⅛ inch or less across, are in long compound clusters. The tiny fruit is covered with spreading hairs.

G. angustifolium is common in Coastal Sage and Chaparral throughout, blooming from May to June.

Although there are four other species of *Galium* in the Santa Monicas, the only other frequent one is *G. nuttallii* which has wider leaves, a smooth fruit and scrambles rather than forms dense mats like *G. angustifolium* does.

The genus has been given the common name of Bedstraw because it was believed that one of its species, *G. verum,* filled the manger of the infant Jesus.

Gala is Greek for "milk." Certain species are used to curdle milk. *Angustifolium* means "narrow foliage."

Thomas Nuttall (1786-1859) apprenticed to his uncle in England, came to Philadelphia as a journeyman printer. On his second day in Philadelphia, he attended a lecture by Benjamin Smith Barton and went on to become a botanist, ornithologist and curator of the Harvard Botanic Gardens. In *Three Years Before the Mast,* he is the character, Old Curious. In 1842, he returned to England as a condition for inheriting an estate from his uncle.

Lizard-Tail Family. SAURURACEAE

This family has been united with the PIPERACEAE in the past and has had Pepper Family as a common name. Our species are perennial, acrid herbs with entire leaves and perfect flowers without petals growing in dense terminal spikes. The calyx when present is often colored like a corolla.

Yerba Mansa

Anemopsis californica Hook.
Lizard Tail Family. SAURURACEAE

Dede Gilman

Yerba Mansa is a perennial herb with numerous basal leaves on long stalks. The stems are hollow and ½ to 2 feet high. The flowers, small in a compact, conical spike surrounded at the base by showy white or pinkish bracts are at the summit of reddish stems. The fruit is a capsule which splits open at the top.

Yerba Mansa is found in very wet soil or shallow water at Point Mugu and Malibu Lagoon. It blooms from April to July.

This plant is much more common in the San Joaquin and Sacramento Valleys than in our mountains. The peppery, astringent, creeping rootstock was prized as a household remedy by early Californians. They chewed the dried root and used an infusion of it for outer aches and pains. The common name, Yerba Mansa, means "tame herb." The true name once was *Yerba del Manso,* "the herb of the tamed Indian."

Anemopsis is from two Greek words meaning "anemone-like."

Saxifrage Family. SAXIFRAGACEAE

Plants in this family are shrubs or herbs, usually with 5 sepals, 5 petals and 2 pistils. Many of them have a cup-shaped base on which the flowers are situated. The leaves are mostly alternate. The fruit is a capsule.

We have five species of the shrubby genus *Ribes* in our mountains and three herbaceous genera. Some botanists separate the shrubs into a family which is named GROSSULARIACEAE. Two other families, HYDRANGEACEAE and PARNASSIACEAE, once included members of the Saxifrage Family.

Woodland Star

Lithophragma affine Gray.
Saxifrage Family. SAXIFRAGACEAE

Geoffrey Burleigh

The Woodland Star is a perennial herb 12 to 18 inches high with slender, flowering stems. The basal leaves are rounded and scalloped. The stem leaves are 3-parted. The 5, small, white petals have 3 lobes. There are 10 stamens on very short stalks.

Woodland Star grows abundantly on shady slopes that are moist in spring and blooms from February to June.

Woodland Star is the only species of its genus in our mountains although there are 6 others in California. It can be differentiated from the similar, but far less common California Saxifrage *(Saxifraga californica)* by the leaves and styles. Woodland Star has leaves on the stems and 3 styles. *S. californica* has only basal leaves and 2 styles.

Another herbaceous genus of this family found in the Santa Monicas is *Boykinia* which has 5 stamens rather than 10 and the stems are covered with long spreading hairs. *B. rotundifolia* and *B. elata* both grow in moist soils near streams but are rarely seen.

Lithophragma is formed from two Greek words meaning "rock" and "hedge or fence." *Affine* means "related."

Golden Currant

Ribes aureum Pursh var. *gracillimum* (Cov. & Britt.) Jeps.
Saxifrage Family. SAXIFRAGACEAE

Dede Gilman

The Golden Currant is a deciduous shrub 3 to 6 feet high with smooth, 3-lobed gray-green leaves. The yellow flowers have a slender tube about ¼ inch long flaring into a flat circle of 5 lobes. They grow in clusters on either side of the main stem. The fruit is a smooth, red or black berry.

Golden Currant is local in Oak Woodland along the north slope of the mountains. It is an early bloomer appearing from January to April.

The genus *Ribes* contains both gooseberries and currants. They are shrubs similar in growth habit and appearance but the stems and fruit of all gooseberries are prickly, and the flowers are elongated tubes. The currants have smooth stems and fruit, and the flowers are open and bell-shaped.

Because of the neat foliage, the bright yellow flowers and colorful berries, Golden Currant is very suitable for ornamental plantings. It propagates easily from suckers or cuttings.

Ribes is from Syrian or Kurdish, *Ribas,* which was derived from an old Persian word. *Aureum* means "golden."

Chaparral Currant

Ribes malvaceum Sm. var. *viridifolium* Abrams.
Saxifrage Family. SAXIFRAGACEAE

Janet Vail

The Chaparral Currant is a shrub 3 to 6 feet high. The ½ to 1¼ inch leaves, dull green above and lighter green and hairy beneath, are 3 to 5-lobed. The rose-pink flowers, in clusters of 10 to 25, grow on either side of the main stem. Each has a slender tube flaring into a flat circle of 5 lobes. The fruit is a dark purple to black smooth berry.

Chaparral Currant is very common in Chaparral up to 2700 feet in elevation throughout. It blooms from October to March.

Dede Gilman

This shrub is very hardy and grows well on dry slopes. It is easily propagated from seeds, cuttings or by layering.

The fruit of all of the *Ribes* species in our mountains is unrewarding, being dry and bitter or insipid. Indians are reported to have used the roots of this one as a remedy for toothache.

Malvaceum means "mallow-like" and refers to the shape of the leaves.

Fuchsia-flowered Gooseberry
Ribes speciosum Pursh.
Saxifrage Family. SAXIFRAGACEAE

Margaret Stassforth

The Fuchsia-flowered Gooseberry is an evergreen shrub 4 to 8 feet high with reddish, bristly branches which are armed with stout, triple spines where the rounded, slightly lobed, dark green leaves meet the stems. The flowers are bright red, ½ inch or more long. The stamens are twice as long as the flower and hang out. The sepals are longer than the petals and resemble them. The fruit is sticky and exceedingly prickly.

Fuchsia-flowered Gooseberry is common throughout away from the immediate coast and blooms from February to April.

The general aspect of the brilliant, abundant, drooping blossoms which line the stems of the Fuchsia-flowered Gooseberry for a space of sometimes several feet suggests a small cultivated fuchsia. A bush in full bloom is a memorable sight. It does well in cultivation.

One species of hummingbird, during a rest stop of several weeks on its annual northward migration, relies entirely upon this plant.

Speciosum is "showy, good looking" which this shrub surely is.

Figwort Family. Snapdragon Family. SCROPHULARIACEAE

Members of the Figwort Family are herbs or shrubs with 5 united sepals and 5 petals with a tubular throat which expands into 2 lips, the upper lips with 2 lobes and the lower lip with 3. The stamens are often in 2 pairs, 1 shorter. The fruit is a dry capsule.

This is a large family familiar in gardens (Snapdragon, Foxglove) and generously represented in the wild in California. There are 13 genera in our mountains, all of them native and many of them showy.

White Snapdragon

Antirrhinum coulterianum Benth. in DC.
Figwort Family. SCROPHULARIACEAE

Dede Gilman

The White Snapdragon is a stout, erect annual 1 to 4 feet tall. The leaves are narrowly oval, ¾ to 3½ inches long. The 2-lipped, white flowers are often tinged with purple and appear densely crowded in terminal clusters on either side of the main stem. These clusters may be up to 10 inches long. The little flowers are shaped exactly like the traditional garden Snapdragon, forming a closed box which requires strength to open on the part of the insect pollinator.

The White Snapdragon occurs in Chaparral, especially on burns, away from the coast. It bloom from April to June.

There are four species of *Antirrhinum* in the Santa Monicas: *A. kelloggii,* a tender sprawling vine (see next page); *A. multiflorum,* a short-lived perennial with rose-red to white flowers; and *A. nuttallianum,* an annual with purple flowers and wide oval leaves.

Antirrhinum, difficult to spell, simply means in Greek, "like a snout" and was bestowed on this genus because the flowers do seem to have snouts. A botanical pioneer of the Pacific Coast, Dr. Thomas Coulter (1793-1843), is honored in the species name. He is credited with being the discoverer of the Colorado Desert in 1831. The Coulter Pine is also named for him.

Twining Snapdragon
Antirrhinum kelloggii Greene.
Figwort Family. SCROPHULARIACEAE

Frances Vogel

The Twining Snapdragon is an annual herb 1 to 3½ feet high with hair-like branchlets beginning where the leaves meet the stems. These little arms wave about until they find support. The oval leaves, ¼ to ¾ inch long, are alternate above and opposite below. The tiny, ½ inch long, blue flowers also wave about on their 3 inch long stems. They are typically snapdragon in shape.

Twining Snapdragon is frequently a fire follower in Coastal Sage and Chaparral throughout the mountains. It blooms from March to May.

This representative of a usually showy genus is hard to spot, but once seen, is easy to recognize.

Dr. Albert Kellogg (1813-1887) was a northern California botanist and one of the seven founders of the California Academy of Sciences.

Indian Paintbrush
Castilleja affinis H. & A.
Figwort Family. SCROPHULARIACEAE

Arline Denny

The Indian Paintbrush is a perennial covered with soft, short hairs. The stems of this species are 12 to 20 inches high, purplish and more or less woody at the base. The leaves have smooth margins and are several times longer than wide, palmately divided into 3 or 5 lobes. The flowers are in a terminal spike subtended by leaf-like, fingered bracts which have scarlet tips, the whole effect being that of a brush dipped into red paint. These fingers are what one sees and takes for flowers. The actual flowers are minute and hardly recognizable as such.

C. affinis grows in open grassy areas and among shrubs in Coastal Sage and Chaparral throughout. It blooms from February to May.

An interesting characteristic of the whole genus is the presence of suckers on the roots enabling the extraction of sustenance from the roots of other plants, an instance of partial parasitism.

We have three other species of this genus in the Santa Monicas, all red-flowered and easily recognized as Indian Paintbrushes. California Threadtorch *(C. stenantha),* found on streamsides, is the only annual. *C. foliolosa* has hairy, whitish leaves and stems. *C. martinii* closely resembles *C. affinis,* but the sepals are shorter and the foliage is sticky.

Castillejo, a professor of botany at Cadiz, Spain, is honored by the generic name. He died sometime before 1781. *Affinis* means "bordering or related to."

Purple and White Chinese Houses
Collinsia heterophylla Buist ex Grah.
Figwort Family. SCROPHULARIACEAE

Margaret Stassforth

Purple and White Chinese Houses are annuals a foot or so high. The long, pointed, upper leaves hug the stems. The lower leaves are oblong. The corolla has 5 lobes, but only 4 are visible. The petals are joined part way to form a tube at the end of which the upper lip stands erect. The 3-lobed, lower lip extends forward; however, the central lobe is folded and entirely concealed by the other two on either side. This is a very intricate arrangement. In this species the upper lip is white and the lower lip is violet. The flowers occur in a series of whorls along the upper part of the stem.

Purple and White Chinese Houses are apt to be abundant when found in Oak Woodland and openings in the Chaparral. They prefer locations which have been moist in the early spring. They bloom in April and May.

The common name arises from the resemblance of its ascending whorls to a pagoda. The generic name commemorates an excellent botanist of the early 1800's, Zaccheus Collins (1764-1831) of Philadelphia. Collins was a wealthy Quaker merchant, a mineralogist and an authority of lower plants, but he was too modest to publish any of his work. *Heterophylla* means that the leaves are different on the same plant. Some leaves are entire and some have rounded teeth.

Dark-tipped Bird's Beak

Cordylanthus filifolius Nutt. ex Benth. in DC.
Figwort Family. SCROPHULARIACEAE

Linda Hardie-Scott

This plant is a hairy, loosely-branched annual 1 to 3½ feet high. The alternate leaves have 3 to 5 thread-like divisions. The floral leaves and bracts are bristly on the margins and generally marked at the tip with a depressed gland. The greenish-yellow or purplish flowers are borne in many-flowered terminal heads. The flowers are two-lipped, ½ to ¾ inch long and nearly hidden within the green calyx. The upper lip encloses the stamens and the style. *Cordylanthus* spp. are hemi-parasitic on various host plants.

C. filifolius is found in open grassy areas in Chaparral and Oak Woodland at high elevations and away from the ocean. It blooms from June to September. Our other species, *C. maritimus,* has simple leaves and is found in the salt marshes of Mugu Lagoon.

The curious tip of the corolla, somewhat suggesting a bird's beak, is responsible for the common name given these grayish-green little herbs. Indians are reputed to have used the plant medicinally as an emetic.

In Greek, *cordule* means "club" and *anthos* "flower." *Filifolius* refers to the thread-like foliage.

Wide-throat Monkey Flower

Mimulus brevipes Benth.
Figwort Family. SCROPHULARIACEAE

Linda Hardie-Scott

The Wide-throat Monkey Flower is a sticky, hairy annual 4 to 30 inches high with unbranched stems. The smooth-margined leaves at the base are oval but the upper ones are narrowly-pointed and cling to the stem. The 2-lipped corolla is lemon yellow, 1 to 2 inches long, with a pair of ridges running down the open throat.

M. brevipes blooms in April and May throughout the mountains away from the coast.

The pale blossoms and the fragile appearance of this Monkey Flower do not lead you to expect to find it in its harsh habitat on dry, rocky hillsides. It is abundant after burns where the contrast between its delicate petals and the black ash surrounding it is most apparent.

Mimulus may come either from the Greek, *mimo,* "an ape," because of a resemblance on the markings of the seeds to the face of a monkey; or, from Latin, *mimus,* "an actor or mimic," because the flower is like the mouth-piece of one of the masks worn by classical actors.

Scarlet Monkey Flower
Mimulus cardinalis Dougl. ex Benth.
Figwort Family. SCROPHULARIACEAE

Dede Gilman

The Scarlet Monkey Flower is a stout, branching, hairy and sticky, perennial herb, 2 to 4 feet high. The opposite, oval leaves are borne directly on the stems. The scarlet, 2-lipped flowers are about 2 inches long with the upper lip very erect and the lower lip turned back. The 4 stamens extend conspicuously from the blossom. The flowers are borne on upright stalks about 3 inches long arising from the axils of the upper leaves.

Scarlet Monkey Flower occurs along streams in Santa Ynez Canyon and westward below 2000 feet. It blooms from May to October.

David Douglas discovered this species and sent the seed to England in the 1830's. The handsome flowers and the distinctive color soon made it popular with florists, and it has continued to be a favorite in cultivation. Its long flowering season makes it desirable but it may become too large and sprawling for the average garden.

Cardinalis which once meant "chief or principal" seems, in this case, to mean "red," the color of the garb of cardinals whether church officials or birds.

Common Large Monkey Flower

Mimulus guttatus Fisch. ex DC.
Figwort Family. SCROPHULARIACEAE

Dede Gilman

The Common Large Monkey Flower is an erect perennial with watery stems 2 to 40 inches high. The oval to rounded leaves, ½ to 3 inches long, are irregularly toothed. The flowers are bright yellow, often spotted with red in the throat, ½ to 1½ inches long on stalks ¾ to over 2 inches long. The fruit is an oval capsule with thin hard walls.

M. guttatus is found in moist areas throughout the mountains and blooms from March to August. The Santa Monicas do not have the showy stands of this species found in moist meadows elsewhere, but it is often seen on any walk during the blooming season.

The Common Large Monkey Flower makes a fine pot plant and with water and small amounts of fertilizer is almost ever-blooming. The pot is soon filled with bending stems which have rooted and can be separated and planted in other pots.

Guttatus is from Latin meaning "a drop-like spot" which describes the red dots on both petals and sepals.

Bush Monkey Flower, Sticky Monkey Flower

Mimulus longiflorus (Nutt.) Grant.
Figwort Family. SCROPHULARIACEAE

Dede Gilman

The Bush Monkey Flower is a much-branched shrub from 2 to 6 feet high. It exudes a gluey substance. The opposite, roundish, sharp-toothed leaves are an inch or so long. The 1 to 2 inch long, light orange to pale, cream-yellow flowers, have a 2-lipped corolla with rough-edged lobes. The calyx is irregularly 5-toothed. The sensitive white lips of the stigma close upon being touched or after receiving pollen. If pollen is actually deposited, the stigma lips remain closed, but they open a short time later if they were fooled. You can play this game with the Bush Monkey Flower using a pin and a bit of patience.

Bush Monkey Flower is quite common in Coastal Sage and Chaparral throughout and may be found in bloom from January through May.

The shrubby growth of this *Mimulus* puts it in a class by itself since all the others in our mountains are herbaceous. Consequently, some botanists prefer the generic name *Diplacus*. The color variations are notable—cream-colored, salmon, even brick-red in the Santa Susana Pass area.

Longiflorus refers to the length of the corolla.

Downy Monkey Flower
Mimulus pilosus (Benth.) Wats.
Figwort Family. SCROPHULARIACEAE

Geoffrey Burleigh

The Downy Monkey Flower is a low annual, 2 to 14 inches high, with many soft hairs on the stems and leaves. The oblong leaves, up to an inch long, and smooth-margined, attach directly to the stems. The small, yellow flowers, only a little over ¼ inch long, have maroon spots at the base.

Downy Monkey Flower is seen occasionally in moist areas in Oak Woodland and Chaparral. It blooms in April and May.

Unlike most of the other members of this genus, this inhabitant of dried stream beds has small pale blossoms and may well be overlooked. A close examination of the flowers makes it plain that this is, indeed, a Monkey Flower.

Pilosus is from Latin and means "hairy."

Owl's Clover

Orthocarpus purpurascens Benth.
Figwort Family. SCROPHULARIACEAE

Dede Gilman

Owl's Clover is an erect, hairy annual growing a foot or more tall with a branched stem bearing many pinnately cleft leaves which are divided into thread-like lobes. The flowers are in a dense spike at the top of each branch. Each flower is 2-lipped. The narrow, upper lip is crimson, the lower lip is creamy-white at the center deepening outwardly to magenta and is inflated in the form of a sac. The bracts and calyx have purplish tips.

While driving through the hillsides of the Santa Monicas, one can catch glimpses of pinkish-red patches of color. More than likely, these are Owl's Clover. It flourishes in open Chaparral and along the coast. It blooms from March to June.

Why this plant should be called Owl's Clover is not obvious. One book suggests that the flower faces have an owlish look. The Spanish Californians called them *Escobitas,* "little whisk brooms," which is quite descriptive.

Like Bird's Beak and Indian Paint Brush, Owl's Clover is partially parasitic.

Orthocarpus means "straight fruit" and *purpurascens* means "becoming purple."

Indian Warrior

Pedicularis densiflora Benth. ex Hook.
Figwort Family. SCROPHULARIACEAE

Linda Hardie-Scott

The Indian Warrior is a stout perennial 6 to 20 inches high. The leaves, bronze when young and becoming green with age, are finely dissected and feather-like; many are basal. The crimson flowers, an inch or so long, are 2-lipped. The upper lip is much more conspicuous than the lower lip; the tube is compressed at the sides and arched above. The flowers are in crowded spikes at the ends of the unbranched stems. The plant is partially parasitic.

Indian Warrior occurs in chaparral clearings throughout the mountains. It blooms from February to April.

Pedicularis means "louse" in Latin and takes us back to an old English belief that when cattle grazed on these plants they became infected with lice. Some people still call the Indian Warrior, Lousewort, wort being an old English word for "plant." *Densiflora* means "densely flowered."

Scarlet Bugler
Penstemon centranthifolius Benth.
Figwort Family. SCROPHULARIACEAE

Janet Vail

Scarlet Bugler is a perennial herb that grows up to 3 feet tall. The stem bears pairs of gray-green, lance-shaped leaves, mostly without stalks, the upper leaves clasp the stem with their heart-shaped bases. The united petals form bright scarlet, tubular flowers an inch or more long with 5 small teeth at the end. The flowers grow in narrow, loose clusters, all tending toward the same side of the stem. As in all the Penstemons, there are 5 stamens and the lowest one has no anther and so does not produce pollen. In *P. centranthifolius,* the sterile stamen has no beard of hairs on it.

Scarlet Bugler is found occasionally in Oak Woodland and Chaparral in our mountains. It blooms in April and May.

The slender flowers suggest trumpets of Honeysuckle on first encounter. Bees visit this species as persistently as they do Honeysuckle. Hummingbirds enjoy them as well and in some areas they have been called Hummingbird's Dinner Horn.

Penstemon means "5 stamens." *Centranthifolius* means "leaves like *Centranthus.*" *Centranthus* is a genus of the Valerian Family, sometimes spelled *Kentranthus.* (See *Kentranthus ruber.*)

Heart-leaved Penstemon, Climbing Penstemon

Penstemon cordifolius (Benth.) Straw.
Figwort Family. SCROPHULARIACEAE

Dede Gilman

The Heart-leaved Penstemon is a straggling vine-like perennial 3 to 9 feet long that clambers over other shrubs. The leaves are heart-shaped, an inch or so long, toothed on the margins, dark green and somewhat leathery. The 2-lipped, orange-red flowers, 1¼ to 2 inches long, droop in the leaf axils. The stamens and pistil are conspicuous in the extended lower lip. Examine the fifth stamen. It resembles a very small golden brush.

Heart-leaved Penstemon is common in Chaparral, Coastal Sage and Oak Woodland throughout the mountains. It blooms from March to August.

This plant has been called Scarlet Honeysuckle. The general look of the flower and the habit of the plant give rise to this misnomer, although the foliage is more suggestive of the cultivated fuchsia.

This is our only shrubby Penstemon. Recently, botanists are calling the shrubby Penstemons, *Keckiella.* David Daniels Keck (1903-), an American botanist known for his work in experimental taxonomy, did *Penstemon* studies and collaborated with Philip A. Munz on *A California Flora.*

Cordifolius means, in Latin, "heart-shaped foliage."

Foothill Penstemon

Penstemon heterophyllus Lindl.
Figwort Family. SCROPHULARIACEAE

Geoffrey Burleigh

The Foothill Penstemon is a slender, many-stemmed, smooth perennial 2 to 5 feet high and woody at the base. The opposite leaves are narrow, lance-shaped and clustered. The funnel-formed, violet flowers, an inch long or more, are 2-lipped and somewhat inflated on one side. The flowers appear at the ends of the stems and branches. The buds are conspicuously yellowish.

Foothill Penstemon is found occasionally in Oak Woodland and Chaparral back from the immediate coast. It blooms from April to June.

The anthers of this species are interesting under a glass—they look like miniature, fringed horseshoes.

This species will produce its lovely flowers in your garden on well-drained slopes and terraces. It requires sun and moderate water.

The species name *heterophyllus* means "having different leaves on the same plant."

Showy Penstemon
Penstemon spectabilis Thurb. ex Gray.
Figwort Family. SCROPHULARIACEAE

Frances Vogel

Showy Penstemon is a stately, handsome, herbaceous perennial from 3 to 6 feet high. The opposite, pale-green, oval, rather leathery leaves have spiny teeth. The upper hairs of leaves join at their broad bases and clasp the stem as though pierced by it. The rose-purple or lilac flowers are 2-lipped, very showy, somewhat inflated at the throat, an inch long, and numerous in compound clusters.

Look for Showy Penstemon on open slopes in Coastal Sage and Chaparral away from the coast when it blooms in April and May.

The genus *Penstemon* is a very large one in North America. There are at least 50 species on the Pacific Coast. Many of them possess flowers of great beauty and have been introduced into gardens.

The dense hairs often appearing on the 5th, or sterile, stamen have caused them to be known as Beard Tongues.

Spectabilis means "spectacular" as this wildflower certainly is.

California Bee Plant, California Figwort

Scrophularia californica C. & S. ssp. *californica*
Figwort Family. SCROPHULARIACEAE

The California Bee Plant is a branching plant with square stems 2 to 6 feet high. The opposite leaves are oval to triangular, 2 to 3 inches long and are coarsely-toothed. The dull, red-brownish flowers are only ¼ inch long appearing in a compound cluster at the top of the stems. The upper lip of the rounded corolla is erect with 4 lobes. The lower lip has 1 lobe and extends outward. There are 4 stamens in 2 pairs and a rudimentary one in the form of a scale on the corolla throat.

California Bee Plant can be found in moist areas throughout and blooms from March until June.

The smallness of the flowers may cause it to be overlooked entirely, but they are alert little flowers with a gnome-like sort of character that rewards their notice. Two of the stamens project just over the lower rim of the corolla like the front teeth of a tiny rodent.

Some species of this genus have rhizomal knobs that were supposed to heal scrofula, a tuberculosis of the lymph nodes, hence the name. In 1474, Matteo Silvatico, an Italian physician, noticed the resemblance between the rhizomal knobs and scrofula and named the plant.

Water Speedwell

Veronica anagallis-aquatica L.
Figwort Family. SCROPHULARIACEAE

Barry Silver

The Water Speedwell is a slightly succulent, perennial herb with stems 4 to 40 inches high. The oblong, opposite leaves appear directly from the stems and are minutely toothed. The many-blossomed flowerstalks arise from the leaf axils. The flowers are pale blue, 4-lobed and ⅛ to ¼ inch across. There are only 2 stamens.

Water Speedwell is found occasionally by lakes and streams in the central part of the mountains. It blooms from May to October.

This is a species originally from the Old World. It is always a delight to discover and examine.

This genus was possibly named for Saint Veronica, because the markings on some species resemble the markings on her sacred handkerchief. Other possible derivations are *Ver*, "the spring," or the Vetonica, a province of Spain. *Anagallis* comes from two Greek words meaning "to delight in again," since the flowers open when the sun strikes them. *Aquatica* refers to the damp areas where the plant is found.

Nightshade Family. SOLANACEAE

Plants in the Nightshade Family are herbs or shrubs with alternate leaves and are usually rank-scented. The flower parts are in 5's, the petals are united. The fruit is a berry or a capsule.

The family is notorious for the number of species that contain poisonous alkaloids, such as atropine and nicotine. A few furnish useful foods—the tomato, the potato and peppers. Tobacco is also a member, and so is the garden favorite, Petunia.

Of the five native genera in the Santa Monica Mountains, you will most often see the Purple Nightshade. Tree Tobacco is even more ubiquitous, but it is not a native.

Toloache, Jimson Weed

Datura meteloides A. DC.
Nightshade Family. SOLANACEAE

Dede Gilman

Datura is a bushy, perennial herb covered with short, gray hairs. The stems, 2 to 4 feet long, are dark green. The coarse leaves are 1½ to 4½ inches long. The showy, fragrant flowers are pale violet to oyster-white, trumpet-shaped, tipped with sharp teeth and up to 10 inches long. The fruit is a prickly, nodding capsule and the large seeds are poisonous.

Datura blooms throughout the year in disturbed soil in low elevations throughout, often along roadsides.

This was a famous plant among the Indians because of its narcotic properties. From the crushed roots, they brewed a liquid which was used in the rites of manhood. It was also used to stimulate their young dancing women. However, deaths have been documented (even recently) from the ingestion of this plant.

Datura stramonium, a native of tropical America, may also be found. It is an annual with smaller flowers and an erect capsule. This species was first encountered in Jamestown in early American history and was the source of its common name.

The roots of the common name, Toloache, are given as *Toloa,* "to nod" and *Tzin,* given in Spanish as "reverential."

Datura comes from the Hindu name *Dhatura. Meteloides* indicates that the plant resembles the *Datura metel* of India.

Indian Tobacco

Nicotiana bigelovii (Torr.) Wats.
Nightshade Family. SOLANACEAE

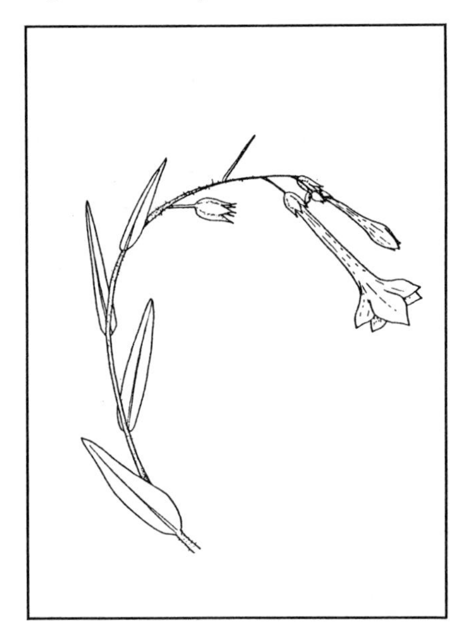

The Indian Tobacco is a sticky, hairy, rank annual 1 to 3 feet high. The leaves are narrowly oval, 2 to 8 inches high. The white, tubular corolla flares into a 5-lobed saucer 1½ to 2½ inches long.

Indian Tobacco is rare in our mountains. When it is found, it is on disturbed soil or burns. It blooms in May and June.

The smoking of tobacco was more common among northern and desert tribes than southern ones. Smoking was a man's activity, unless a woman was the shaman. The cured leaves were hung in little baskets. Pipes were made of wood or soapstone.

Nicotiana commemorates Jean Nicot (1530-1600). Nicot was the French Ambassador to Portugal (1559-1561) and introduced tobacco into France about 1560. He also wrote one of the first French dictionaries. This species is named for Bigelow. (See *Coreopsis bigelovii.*)

Tree Tobacco

Nicotiana glauca Grah.
Nightshade Family. SOLANACEAE

Dede Gilman

Tree Tobacco is a straggling shrub, or small tree, from 6 to 20 feet high. The ample, bluish-green leaves are oval, very smooth with a "bloom" which can be rubbed away. The long, greenish-yellow tubular flowers, 1½ to 2 inches long, appear loosely at the ends of the branches.

Tree Tobacco is common everywhere especially in disturbed areas and along stream beds. It may be found blooming all year.

This plant was introduced from South America in Spanish days. One version is that it arrived here when the padres imported grain. Another has it that it was purposely brought by the padres as a source for smoking tobacco. It is poisonous to ingest either cooked or raw.

Glauca is from Greek and means "bluish-gray," referring to the leaves.

White Nightshade

Solanum douglasii Dunal in DC.
Nightshade Family. SOLANACEAE

Geoffrey Burleigh

The White Nightshade is a slightly woody perennial 3 to 6 feet high with scrambling stems. The oval leaves with wavy margins are ¾ to 4 inches long. The small, saucer-shaped, white flowers, about ½ inch across, are 5-parted. The 5 stamens converge into a cone of yellow anthers and project forward. The fruit is a black berry.

White Nightshade is common throughout on brushy slopes and disturbed ground, blooming in February and March.

The flowers of this species are small and unexciting. The plant itself is far from handsome, but the 5-lobed, saucer blossom with fused anthers makes it easy to recognize as a *Solanum.*

Indians are said to have used the juice of the berries for tattooing and to cure inflamed eyes.

Solanum is "quieting" in Latin and was given because of the narcotic properties of some species. This is one more plant named for David Douglas. (See *Artemisia douglasiana.*)

Purple Nightshade
Solanum xantii Gray.
Nightshade Family. SOLANACEAE

Dede Gilman

The Purple Nightshade is a perennial herb several feet high with thin, oval leaves, 2 inches long and sometimes lobed at the base. The flowers are saucer-shaped, about an inch across, deep violet with white encircled green spots in the center. A cone of yellow anthers surrounds the style. The fruit is a purplish berry about ¼ inch in diameter.

Purple Nightshade is very common in Coastal Sage and openings in Chaparral throughout and blooms from January to May.

Five of the seven *Solanum* species in the Santa Monicas are introduced, but *S. xantii* is a native. Its clusters of violet flowers are very handsome. It is reported that *S. xantii* smells like the Wild Rose.

Xantus Janos (a.k.a. John Xanthus) (1825-1894) was a Hungarian who collected in California and Baja California. While at Ft. Tejon, he had the lowly rank of hospital steward and so used the name Xantus on his U.S. military records rather than his family name, de Vesey.

Nettle Family. URTICACEAE

The Nettles are annual or perennial herbs. Most have stinging hairs. The leaves may be either alternate or opposite. The greenish flowers on short strings beneath the leaves are without petals and often are either male or female. The male flowers have 4 sepals, 4 stamens and a rudimentary pistil. Female flowers have 4 unequal sepals and a simple ovary bearing one seed. The fruit is an achene enclosed by the sepals which do not fall.

We have five species of this family in our mountains. There are two without stinging hairs—Pellitory *(Parietaria floridana),* seldom found, and Baby Tears *(Soleirolia soleirolii),* an escaped garden plant.

Western Nettle

Hesperocnide tenella Torr.
Nettle Family. URTICACEAE

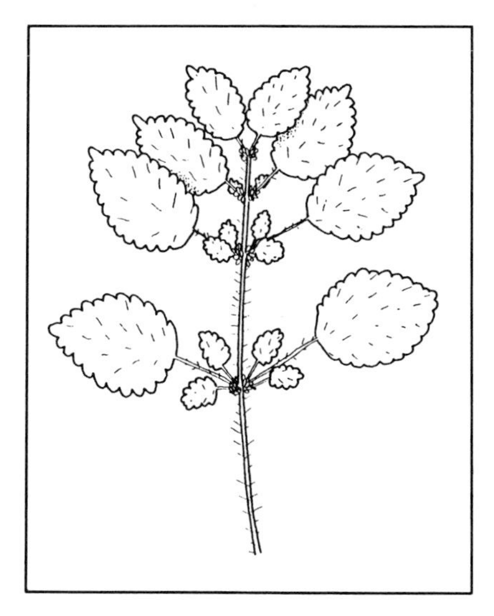

The Western Nettle is an annual herb with slender, weak stems 1 to 2 feet high, that are bristly with scattered hairs. The opposite leaves, ½ to 2 inches long on thin stalks, have toothed margins. The flowers lack petals and are densely clustered at the base of the leaves. The sepals of the female flowers are joined in a tube.

Western Nettle is found on moist slopes in Chaparral and Oak Woodland below 2000 feet and away from the ocean. It blooms in April and May.

In addition to Western Nettle there are two other stinging nettles in the Santa Monicas. Creek Nettle *(Urtica holosericea)* is a perennial. Dwarf Nettle *(U. urens)* is an annual that much resembles Western Nettle. It is naturalized from Europe.

Direct contact with any of the Nettles is to be avoided as the sting contains formic acid and the sensation is that of being stung by many ants. Cooking removes this unpleasant feature and the leaves of young nettles, gathered while wearing heavy gloves, can be substituted in any recipe calling for spinach or chard.

Two Greek words, *hespera,* "west," and *knide,* "nettle," provided the genus name. *Tenella* is from Latin meaning "quite delicate."

Valerian Family. VALERIANACEAE

Valerians are herbs or shrubs with opposite leaves. The flowers are small, 5-lobed, tubular and usually spurred. The fruit is an achene.

Red Valerian

Kentranthus ruber (L.) DC.
Valerian Family. VALERIANACEAE

Linda Hardie-Scott

The Red Valerian is a smooth, perennial herb with several stems up to 3 feet tall, growing from the base. The oval, smooth-edged leaves, 2 to 4 inches long, rise directly from the main stem. The crowded flowers are red to pink and occasionally white.

This native of the Mediterranean has naturalized in moist areas throughout. It blooms from May to August.

Only one other member of this family grows in the Santa Monica Mountains and it is seldom found. *Plectritis ciliosa,* a native, grows in partial shade in the western portion of the range and blooms much earlier than the Red Valerian.

Kentranthus comes from two Greek words meaning "a flower with a spur." *Ruber* is from Latin meaning "red."

Vervain Family. VERBENACEAE

Members of the Vervain Family are herbs, shrubs or trees. Often these plants have square stems. The leaves are opposite. There are 5 or 4 sepals and the petals are united. The 4 stamens, most often paired, are borne on the petals. The fruit can be a few-seeded drupe or 4 nutlets.

Lantana and Verbena of the garden belong to this family, as well as Teak and many other exotic plants of the tropics.

Vervain, Verbena

Verbena lasiostachys Link.
Vervain Family. VERBENACEAE

Sarah Thomas Schwaegler

Vervain is a much-branched, erect perennial, 1 to 3 feet high, with many long, soft hairs. The opposite leaves, $^{3}/_{4}$ to $2^{1}/_{4}$ inches long, are coarsely toothed. The calyx is 5-toothed. The dark blue to blue-purple flowers, in narrow terminal spikes 2 to 8 inches long, are 5-lobed, slightly irregular and less than $^{3}/_{16}$ inch across.

Vervain blooms from April to October, prefers disturbed, moist places and is scattered throughout the mountains.

There is another native *Verbena, (V. robusta),* which is similar to *V. lasiostachys,* but it is seldom encountered. An examination of the nutlets is necessary for differentiation. There are two species of the rarely seen *Lippia* which sprawls rather than being erect, exhibits flowers in globes, not spikes, and has a 2-lobed calyx.

Verbena is the ancient Latin name of the common European Vervain. But, J. Pitton de Tournefort (1656-1708), the greatest continental botanist of his century, believed *Verbena* was a corruption of *Herbena,* i.e., *herba bona,* "the good plant," because it was "in use among the heathens. . . in their religion and worship."

Violet Family. VIOLACEAE

Our native Violets are low perennial herbs. The flowers have 5 petals. The lower petal has an expanded tip which serves as a landing place for insects. There are 5 stamens. The lower 2 stamens have hanging nectaries which curve into a hollow spur at the back of the lower petal.

Most of the members of the Violet Family are tropical shrubs. There are even some trees. Our only species is pictured here.

Johnny-jump-up

Viola pedunculata T. & G.
Violet Family. VIOLACEAE

Janet Vail

The Johnny-jump-up is a slender-stemmed perennial, 4 to 14 inches high. The mostly basal, ¼ to 1¼ inches long heart-shaped leaves have coarsely-rounded teeth on the edge. There are 5 sepals, 5 stamens and 5 yellow petals. The 2 upper petals are tinged with brown outside. The 3 lower ones are veined with purple. The lowest petal has a spur at the base but it is such a short one you must peer for it. The blossoms are borne on naked flower stems 5 or 6 inches long that stand high above the leaves.

Johnny-jump-ups bloom from February to April in grassy areas in Chaparral and Oak Woodland below 2000 feet. They are scattered throughout.

This flower is sometimes called Yellow Violet. The early Californians called them *Gallitos,* "little roosters." The larvae of the Silverspot Butterfly *(Speyeria)* feed on these plants.

Viola is a classical name like *Rosa.* Roses and violets have been loved and named since ancient times. *Pedunculata* points to the prominent flower stems.

Mistletoe Family. VISCACEAE

Mistletoes are evergreen shrubs or herbs that are parasitic on trees or shrubs, absorbing food from the sap of the host through specialized roots. The branched stems are swollen at the joints. The opposite leaves are thick, leathery, without stalks and smooth-margined. Male and female flowers are borne on separate plants. The flowers are 2 to 5-parted, small, greenish and lack petals. There are 2 to 5 stamens and 1 stigma. The fruit is a berry.

There are about 11 genera and 450 species in all continents. We have only one genus and two species in the Santa Monica Mountains.

Bigleaf Mistletoe

Phoradendron tomentosum (Engelm. ex Gray) ssp. *macrophyllum* (Engelm.) Wiens.
Mistletoe Family. VISCACEAE

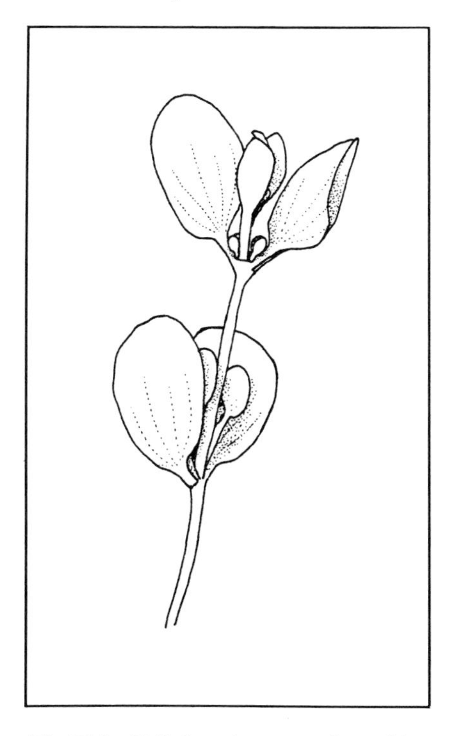

The Bigleaf Mistletoe is a green plant with a brittle woody stem and thick, fleshy oval leaves with a rounded tip. The flowers are inconspicuous, lack petals and are either male or female on separate plants. The fruit is a small white berry less than ¼ inch in diameter. The seeds are covered with a sticky substance which permits them to adhere to the branches of host trees.

This species is found occasionally in the inland portion of the range wherever host trees appear. The blooming period is from December to March.

Bigleaf Mistletoe is chiefly parasitic on Western Sycamore *(Platanus racemosa),* but also may be found on Poplar (*Populus* spp.), Willow (*Salix* spp.) and California Walnut *(Juglans californica).* Hairy Mistletoe *(Phoradendron villosum),* whose leaves are covered with dense soft hairs, grows on Oaks (*Quercus* spp.).

Both species are collected and sold at Christmas, although they are not as ornamental as those gathered in the east.

Phoradendron in Greek means "tree thief." *Tomentosum* means "densely covered with matted wool."

Caltrop Family. ZYGOPHYLLACEAE

Although most members of this family are shrubs or subshrubs with jointed branches, a few are herbs or trees. The leaves are opposite and compound. The solitary flowers have 5 sepals, 5 petals and 10 stamens. When the fruit is dry, it splits into sections.

One well-known member of this family is the Creosote Bush *(Larrea tridentata)* of the Mojave and Colorado Deserts.

Puncture Vine

Tribulus terrestris L.
Caltrop Family. ZYGOPHYLLACEAE

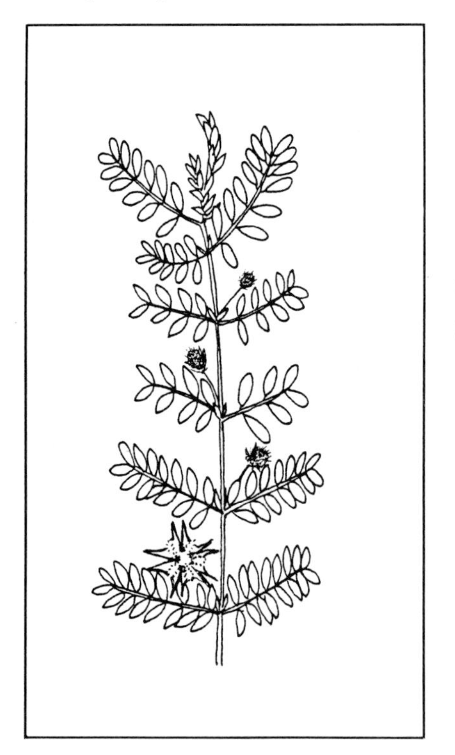

Puncture Vine is a bright green annual. Commonly prostrate, the stems branch freely from the crown. A taproot produces a network of fine rootlets which take advantage of soil moisture, enabling this plant to live under dry conditions that few others can survive. The leaflets of the compound leaves are oppositely arranged. Stems and leaves are densely covered with silky hairs. The bright yellow flowers, borne in the axils of the leaves, are only ⅛ inch long with 5 petals, 5 sepals and 10 stamens. The fruit consists of 5 spiny nutlets or burs.

Puncture Vine appears in waste places throughout (even in your own driveway if you are unlucky). It blooms from April to October.

The five nutlets equipped with 2-4 spreading spines fall apart at maturity pointing one spine upward to be disseminated by animals, shoes or rubber tires. Within the bur, the seeds lie one above the other, separated by hard, horny tissue. The largest seed germinates first and if there is only enough moisture for one, the others remain dormant until a more favorable time.

This introduction from the Mediterranean has become a weed of major concern, damaging cattle, sheep, horses, swine, and dogs and lowering the grade of wool. It is the bane of the bicycle rider.

Tribulus is Latin for "three-pointed, a caltrop." A caltrop is a military weapon (an iron ball with four projecting spikes) used to delay advancing mounted and unmounted troops. *Terrestris* is Latin for "on land."

SOME SUGGESTED WILDFLOWER TRIPS

PARKS

STATE OF CALIFORNIA, DEPARTMENT OF RECREATION AND PARKS

(818) 706-1310. State Parks is by far the largest landowner in the Santa Monica Mountains National Recreation Area with five major parks, the Backbone Trail and a multitude of smaller sites. Most of the day use areas charge a parking fee. Camping reservations for Pt. Mugu State Park, Leo Carrillo State Beach and Malibu Creek State Park can be made through Ticketron. Reservations for trail camps are made at the park.

POINT MUGU STATE PARK. (818) 706-1310. You reach this park by traveling west on Pacific Coast Highway. The entrance station can supply you with small maps guiding you to trails in 16,000 acres that encompass beach, rocky shore, sycamore savannas, grassy uplands and mountain stretches.

The Big Sycamore Canyon Trail is a leisurely one, crossing the stream several times. The many Sycamores are most magnificent.

La Jolla Valley, west of Big Sycamore Canyon, is unique in natural values and scenic beauty—a rolling grassy plain surrounded by mountains and studded with oaks. The La Jolla Trail is steep and narrow, but the trip is worth the effort. The view is spectacular with Boney Mountain, the highest promontory of the range, in the background.

The Serrano Valley Trail parallels Serrano Creek through a woodsy canyon to reach a high mountain grassland.

The trails in the northern portion of the park, filled with sights and surprises, are accessed through Rancho Sierra Vista/Satwiwa, a National Park Service unit in Newbury Park.

MALIBU CREEK STATE PARK. (818) 706-8809, (818) 706-1310. The park entrance is 0.2 miles south of Mulholland Highway on Las Virgenes/Malibu Canyon Road. This is the former 20th Century Fox Movie Ranch, adding historical interest to the natural beauty. Malibu Creek, dammed at the turn of the century, forms the 7-acre Century Lake.

The 4,000 acres that make up the park also encompass ranches once belonging to Bob Hope and Ronald Reagan. There are 20 miles of trail to explore on foot or horseback, some of them affording views of unreachable vertical peaks and rocky outcroppings. The best wildflowers are found in the oak woodlands south of Mulholland Highway in the northwest area of the park, along the road to the lake, and along Las Virgenes Creek in the northern Liberty Canyon area.

POINT DUME STATE BEACH (Westward Beach). Twelve public beaches extend along the 47 miles of Pacific Coast Highway between Santa Monica and Pt. Mugu. To reach Point Dume State Beach, drive west along Pacific Coast Highway 10½ miles beyond Malibu Canyon Road. Turn toward the ocean on Westward Beach Road just *before* the entrance to Zuma Beach. On the bluffs away from the ocean and extending down to the 3 miles of wide sandy beach are a generous assortment of the native plants of the Coastal Sage and Dune associations.

TOPANGA STATE PARK. (213) 455-2465, (818) 706-1310. From Topanga Canyon Boulevard, turn east on Entrada, turning left at each intersection where this is a choice until you arrive at Trippet Ranch, the park headquarters. A display map or the ranger on duty may be able to help you. In the nearly 8,000 acres of Chaparral, Oak Woodland and Grassland, there are 32 miles of trails. The one which is recommended for the lushest displays of wildflowers is the 3½ mile Musch Ranch Trail loop. An active group of docents conducts public walks during the spring flowering, and there is an informative brochure accompanying a well-marked nature trail.

Trails lead into the park from Los Liones, Santa Ynez Canyon, Temescal Canyon, Will Rogers State Historic Park, Caballero Canyon, Serrania Park, Mulholland Highway, and from the parking lot on Entrada 0.1 mile from Topanga Canyon Boulevard.

SANTA MONICA MOUNTAINS CONSERVANCY

(213) 620-2021. This state agency was established to acquire lands and administer programs on an interim basis until the lands could be purchased by the National Park Service or State Parks. Its current authorization expires in 1990.

PETER STRAUSS RANCH. (818) 706-8380. This park, formerly known as Lake Enchanto and most recently owned by Peter Strauss, is on Mulholland Highway west of Malibu Creek State Park. Flowers are abundant along the streambed and along the short loop trail through the oak woodlands. Picnic grounds and a lawn invite the visitor to linger at this enchanting site.

STUNT RANCH. The Stunt High trailhead is 1 mile from Mulholland on Stunt Road. The trail follows Cold Creek riparian area before turning to cross the meadow to Stunt Ranch. The trail continues, jogging back across Stunt Road, twice, before joining the Backbone Trail near Saddlepeak. Variety and abundance of flowers typify this trail.

MULHOLLAND CREST, lower ZUMA CANYON, lower SOLSTICE CANYON, ARROYO SEQUIT RANCH, DEER CREEK RANCH, WILACRE and FRYMAN CANYON are, as of this writing, owned by the Conservancy. Several other wonderful properties are under negotiations. These are all splendid areas sure to have wonderful flower displays, so check with the National Park Service for a status report.

SANTA MONICA MOUNTAINS NATIONAL RECREATION AREA

National Park Service, (818) 888-3770.

CASTRO CREST. The trailhead is 5.4 miles north of Pacific Coast Highway on Corral Canyon Road. Open every day for hiking. Trails and the Castro Peak Fireroad to the west go to upper Solstice Canyon, Castro Peak, Malibu Creek State Park and points beyond (the NPS portion of the Backbone Trail). Mesa Peak Motorway to the east is the state portion of the Backbone Trail that continues to Will Rogers State Historic Park via Tapia Park and Topanga State Park. The Backbone Trail when completed will stretch 55 miles along the crest of the Santa Monica Mountains from Griffith Park to Point Mugu.

PARAMOUNT RANCH. The park entrance is on Cornell Road north of Mulholland Highway. Trails, a riparian area and Western Town, a remnant of movie days, add interest to this park.

ROCKY OAKS. The entrance is on Mulholland Highway, 50 yards west of Kanan Road. A day use area is open for hiking and picnicking. Ranger and docent-led hikes available.

LOS ANGELES COUNTY DEPARTMENT OF RECREATION & PARKS, AND SANITATION DISTRICT

TAPIA COUNTY PARK. (818) 346-5008. Take Malibu Canyon/Las Virgenes Road north from Pacific Coast Highway for 5½ miles. Immediately after Piuma Road, there is a plainly marked entrance to the park. This is a preserve of old oaks and new sewage. On the southern edge of this county park's shady picnic area is a sewage treatment plant accused of polluting Malibu Creek. Critics want the plant relocated. There are a number of easily accessible species of wildflowers along the stream and under the trees. A trail that connects Tapia to Malibu Creek State Park has a variety of flowers and shrubs.

CHARMLEE NATURAL AREA PARK. (213) 457-7247. Enter this county park on Encinal Canyon Road 4 miles north of Pacific Coast Highway. Charmlee has been described as flower patches surrounded by typical Santa Monica Mountains Chaparral. Its oak forest and open meadows cover 450 acres. A nature center, group campground and a docent program is planned.

SULLIVAN CANYON, Pacific Palisades. From Sunset Boulevard, go north 0.2 mile on Mandeville Cyn. Rd. Turn left on Westridge Road, left on Bayliss and left on Queensferry and park near the wooden road barrier. Walk down to the fence. The access trail is along the right side of the fence. The canyon bottom trail winds through a narrow valley whose steep sides restrict the loss of moisture, encouraging abundant plant growth under the shade of Oaks, Sycamores and Walnuts. Under this leafy cover, Sullivan Canyon makes a particularly pleasant late afternoon, late summer of early fall hike, although at this time of the year, the stream is usually dry and the wildflowers are likely to be restricted to fall-blooming species of the Sunflower Family.

SCENIC DRIVES

MULHOLLAND SCENIC CORRIDOR. This scenic drive winds 50 miles west from the Hollywood Freeway to the Pacific Ocean at Leo Carrillo State Beach. A 7-mile portion in the city is not paved. Five overlooks built by the Santa Monica Mountains Conservancy offer spectacular views. Portions pass through Topanga State Park, Malibu Creek State Park and Leo Carrillo State Beach.

Other crest and canyon roads to be considered in the ranch include Potrero Road (through Hidden Valley), Decker Canyon Road, Yerba Buena, Corral Canyon, Encinal Canyon Road, Piuma Road, Stunt Road, Schueren and Las Tunas Canyon Road, all with scenic values and roadside flowers in the spring.

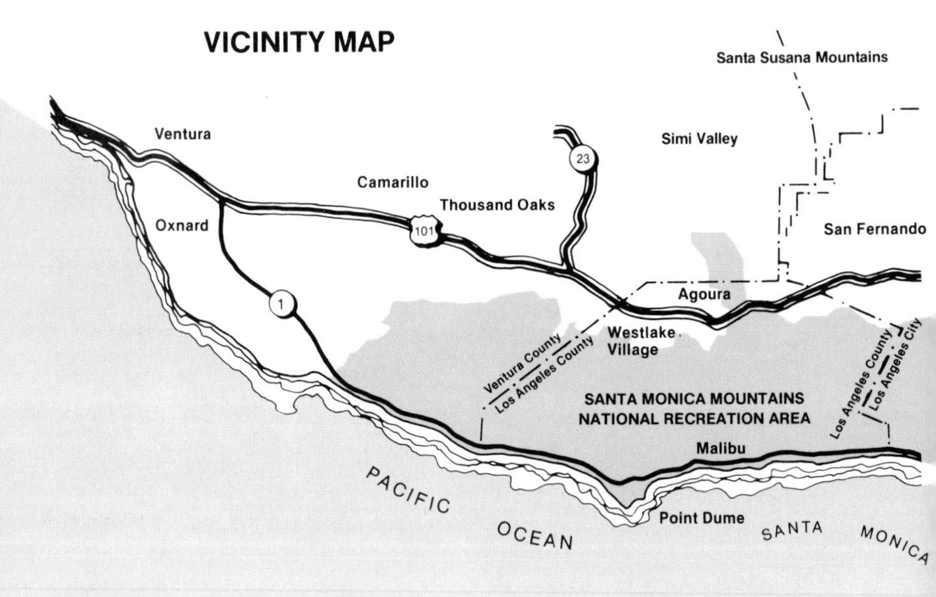

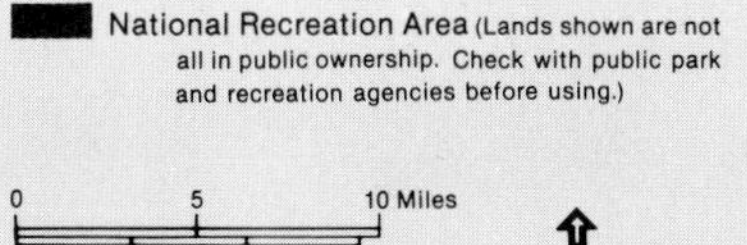

Getting There

At present, private vehicles remain the primary means of access to the Santa Monica Mountains. An RTD bus #175 extends along the Pacific Coast Highway westward as far as Trancas Canyon and provides limited access to the Santa Monica Mountain beaches. For now, visitors must rely on their own vehicles to reach the interior of the mountains.

For Your Information

An information center for the recreation area is located in the National Park Service headquarters at 22900 Ventura Boulevard, Woodland Hills, CA 91364. Phone 818-888-3770 voice or TDD for hearing impaired. Some recreation sites are accessable to wheelchairs. Contact individual parks for details.

Visitor Safety Information

In the mountains: Fire is a constant danger, particularly in the late summer and fall. During high fire danger, some park areas may be closed. Don't build ground fires and use caution even when cooking on grills or on a camp stove. Smoking is prohibited in some areas. Though fire is a normal element of the chaparral ecosystem, accidental fires in populated mountains can endanger lives and destroy millions of dollars in property. Please be careful.

Vehicles, including motorcycles, must stay on paved roads. They are not allowed on trails.

Equestrians can use trails and fire roads on public lands but are not allowed on state beaches. Dogs must be leashed in campgrounds and picnic and parking areas. Dogs are not allowed on trails, beaches or in backcountry park areas. State park regulations require proof of your dog's rabies vaccination.

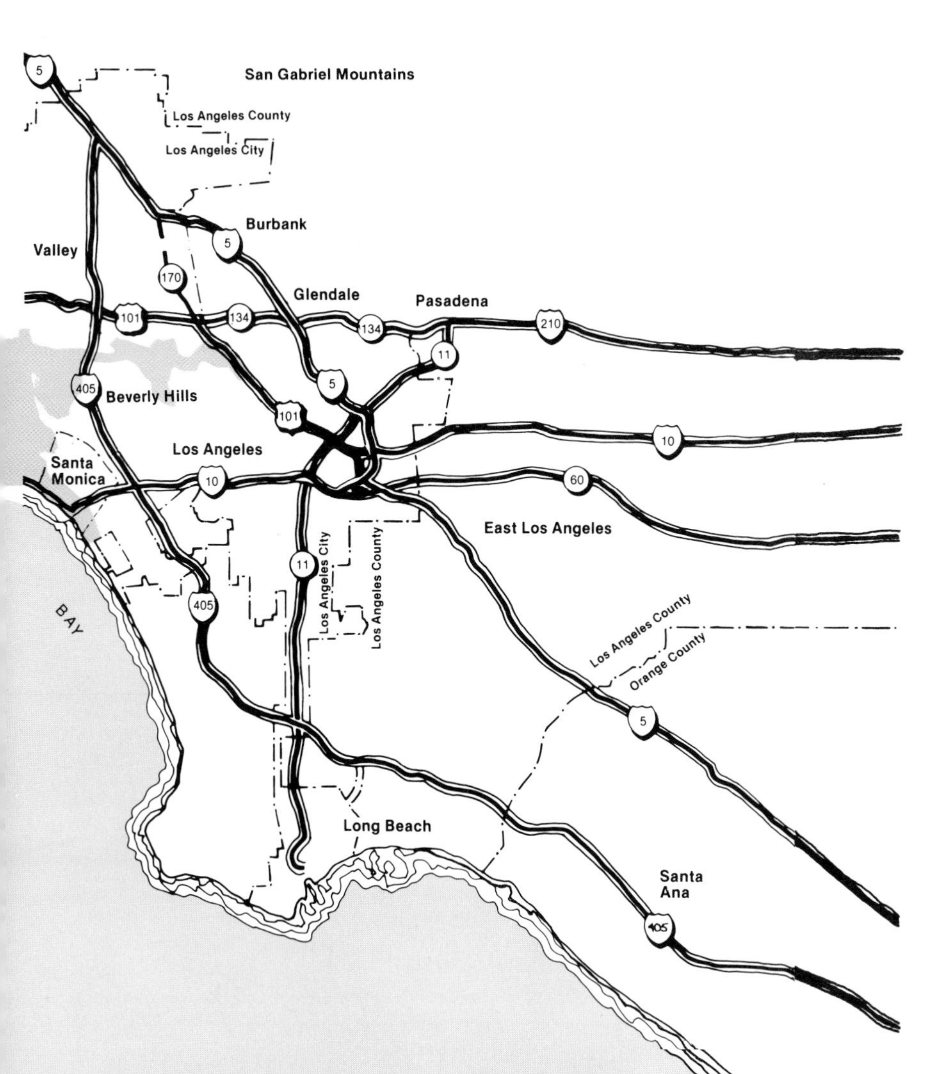

Parks within the Santa Monicas are surrounded by private property. Please respect the rights of our neighbors by staying on public lands.

Plants and animals are protected here. Poison oak grows in most of the shady woods and canyons of the mountains. Learn to recognize this vine with shiny green or red leaves in clusters of three so that you'll be able to avoid it. Rattlesnakes are protected and normally will not strike unless disturbed. Pay attention to where you put your feet and hands. Do not rush through off-trail bushes.

Loaded firearms are not allowed in public parklands.

Litter and trash spoil the mountains for all who follow you. If you brought it in, please take it out.

On the beach: Riptides or undertows flow *away* from shore. If you become caught in a riptide, don't panic. Swim parallel to shore until you are in the surf line, then swim in. Riptides are particularly dangerous at Zuma Beach.

Shuffle your feet when wading in shallow water and you'll warn stingrays of your presence. Jellyfish can sting, too; be careful. Wear tennis shoes when walking on rocky shores to avoid sea urchin spines and sharp edged barnacles.

Do not use glass containers on the beach. Broken glass and pop-tops hidden in sand can cause serious cuts.

Don't fish within 50 feet of swimmers or surfers.

And last of all, swim or dive with a friend.

SANTA MONICA MOUNTAINS

LEGEND

PARKLANDS

1 Point Mugu State Park (DRP)
  2 La Jolla Canyon
  3 Big Sycamore Canyon
  4 Serrano Valley
  5 North Point Mugu through Rancho Sierra Vista
6 Rancho Sierra Vista (NPS)
7 Los Robles Trail (CPR)
8 Deer Creek Ranch* (SMMC)
9 Yellow Hill* (NPS)
10 Leo Carrillo State Beach (DRP)
11 Nicholas Flats (DRP)
12 Malibu Springs (NPS)
13 Arroyo Sequit Ranch* (SMMC)
14 Flora Hill* (NPS)
15 Trancas Canyon* (NPS)
16 Charmlee Natural Area Park (LAC)
17 Decker Canyon Camp (LA City)
18 Zuma Canyon* (NPS)
19 Lower Zuma Canyon* (SMMC)
20 Pt. Dume Reserve (DFG)
21 Solstice Canyon* (SMMC)
22 Castro Crest (DRP, NPS)
23 Rocky Oaks (NPS)
24 Peter Strauss Ranch (SMMC)
25 Paramount Ranch (NPS)
26 Cheeseboro Canyon* (NPS)
27 Malibu Creek State Park (DRP)
  28 Main gate and campground
  29 Reagan Ranch
  30 White Oak Farm*
31 Tapia County Park (LAC)
32 Piuma Trailhead* (DRP)
33 Malibu Canyon Overlook* (SMMC)
34 Malibu Bluff* (LAC, DRP)
35 Malibu Lagoon State Beach (DRP)
36 Malibu Canyon* (DRP)
37 Stunt High Trail to Stunt Ranch (UC, MRT, SMMC)
38 Cold Creek Preserve (MRT)
39 Topanga Meadows* (DRP)
40 Camp Slauson* (SMMC)
41 Topanga State Park (DRP)
  42 Trippet Ranch
  43 Los Liones Trailhead
  44 Santa Ynez Trailhead
  45 Caballero Canyon Trailhead
46 Serrania Park (LA City)
47 Mulholland Crest* (SMMC)
48 Temescal Canyon (SMMC, DRP)
49 Will Rogers State Historical Park (DRP)
50 Sullivan Canyon* (LAC Sanitation District)
51 Cross Mountain Park
  52 Franklin Canyon (NPS, WODOC [William O Douglas Outdoor Classroom])
  53 Wilacre Park (SMMC)
  54 Tree People
55 Fryman Canyon* (SMMC)
  56 Fryman Overlook (SMMC)
  57 Universal City Overlook (SMMC)
  58 Runyon Canyon* (LA City)
  59 Hollywood Bowl Overlook (SMMC)
  60 Griffith Park (LA City)

BEACHES

61 Mugu Rock Beach
62 La Jolla SB (Campground/day use) (DRP)
63 Sycamore Cove/Big Sycamore Cyn Campground (DRP)
64 Leo Carrillo SB
  County Line Beach
  Staircase Beach
  Campground and day use
65 Nicholas Canyon County Beach
66 El Pescador SB
67 La Piedra SB
68 El Matador SB
69 Zuma County Beach
70 Westward Beach
71 Pt. Dume SB
72 Paradise Cove
73 Dan Blocker SB
74 Corral SB
75 Malibu Lagoon SB
76 Malibu Surfrider SB
77 Malibu Pier (DRP)
78 Las Tunas SB
79 Topanga SB
80 Will Rogers SB
81 Santa Monica Pier

82—DRP—State of California Department of Recreation and Parks (818) 706-1310, (805) 987-3303, (213) 454-8212
83—NPS—National Park Service (818) 888-3770
LAC—Los Angeles County Parks and Recreation Dept. (213) 457-7247, (805) 259-7721
LA City—Los Angeles City Recreation and Parks Dept. (213) 485-5515
CPR—Conejo Parks and Recreation (805) 495-6471
SMMC—Santa Monica Mountains Conservancy (213) 620-2021. SMMC properties will be turned over to DRP or NPS.
MRT—Mountains Restoration Trust (213) 456-5625
DFG—Department of Fish and Game (213) 590-5132
UC—University of California

*Publicly owned parklands, some with trails but little or no other facilities as of January 1986.

Backbone Trail — — — — — — — —

Backbone Trail (proposed) • • • • • •

Backbone Trail from Will Rogers to Malibu Creek State Park (DRP) is mostly operational. Call (818) 706-1310 for status.

Backbone Trail from Malibu Creek State Park to Pt. Mugu State Park (DRP, NPS). Portions are operational. Call (818) 888-3770 for update on status.

Mulholland Scenic Corridor. This 55-mile drive starts east of the Hollywood Freeway and continues 55 miles westerly to Leo Carrillo State Beach at Pacific Coast Highway. A 7-mile city portion is unpaved.

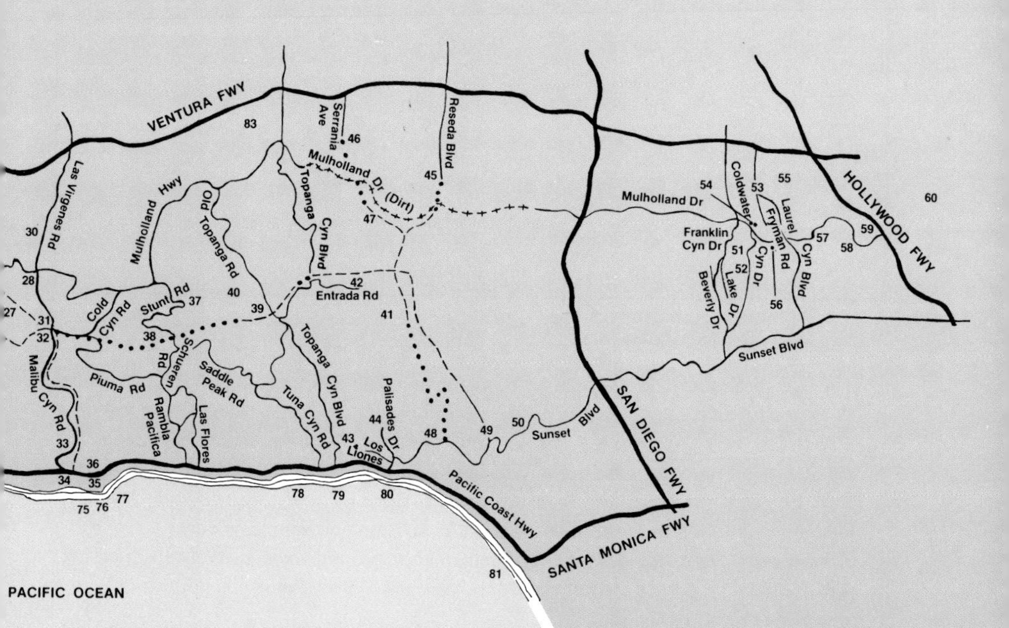

Since the Santa Monica Mountains National Recreation Area is still in the acquisition and development phase, this map represents ownerships and road alignments as of January 1986. Updated maps can be acquired at the National Park Service office.

NATIVE TREES OF THE SANTA MONICA MOUNTAINS

Chaparral, that unique plant community indigenous to our Mediterranean climate, has been called by some "the Elfin Forest," indicating that the area, covered as it is by many bushes and few trees, is not a true forest. The trees of the Santa Monicas, while not crowded together or numerous, are well worth learning to recognize.

Perhaps most typical and most often encountered are the members of the Beech Family, the FAGACEAE. The four species reported all produce dangling male catkins and acorns enclosed in scaly bracted cups. They hybridize readily which makes identification sometimes difficult. All are in the genus *Quercus.*

Q. dumosa is the only shrub. But, since trees have a shrubby youth, start your identification by deciding if the scales on the acorn cups are thin and papery, or thick and warty. Scrub Oak *(Q. dumosa)* and Valley Oak *(Q. lobata)* have warty scales. Scrub Oaks have leaves ¾ inch to 2 inches long which, when cleft (not all of them are), are sharply toothed. The leaves of the Valley Oak are 2 to 4 inches long and the lobes are rounded. The Valley Oak is the only deciduous one of the four species.

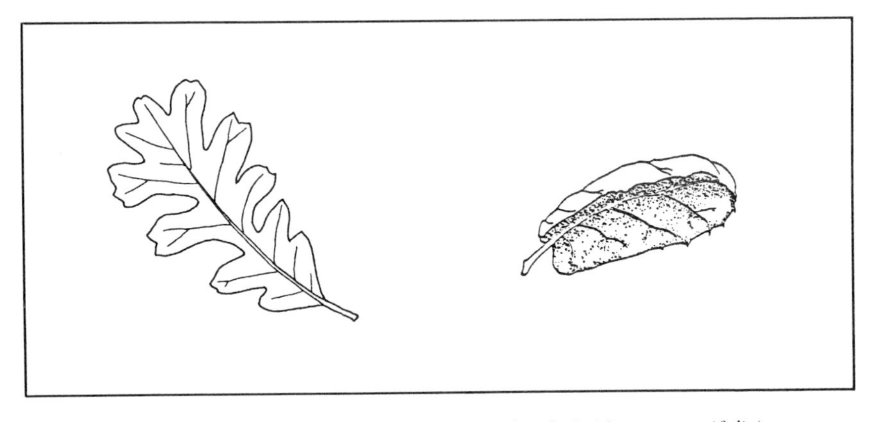

Valley Oak *(Quercus lobata)* Coast Live Oak *(Quercus agrifolia)*

The two species with thin papery scales on the acorn cups, Coast Live Oak *(Q. agrifolia)* and Interior Live Oak *(Q. wislizenii),* are best differentiated by their leaves. The leaves of the Coast Live Oak are convex and show tufts of hair on the veins on the underside. The leaves of the Interior Live Oak are flat and hairless. The Interior Live Oak is only in the Triunfo Lookout area, Saddle Peak and Santa Ynez Canyon in the Santa Monicas, although it is common in Foothill Woodlands as far south as Ventura County.

Acorns were the staple diet of the Indians. They ground the acorns when dried and leached the meal to remove the tannic acid. It is estimated that it took 500 pounds of acorns for one family each year.

When you see our one species of PLATANACEAE, Western Sycamore *(Platanus racemosa),* you have also discovered a stream bed or at least an old stream bottom. Our Sycamores can be tall, 40 to 90 feet, and erect. More often though, they have large, crooked trunks and branches which nearly touch the ground. The bark near the base of the trunk is dull brown and quite ridged, but slightly further up the trunk, the bark is smooth and ashen in color with mottled greenish patches. The 3 to 5-lobed leaves are up to 10 inches long and 12 inches wide. The fruit is a characteristic bristly "button ball."

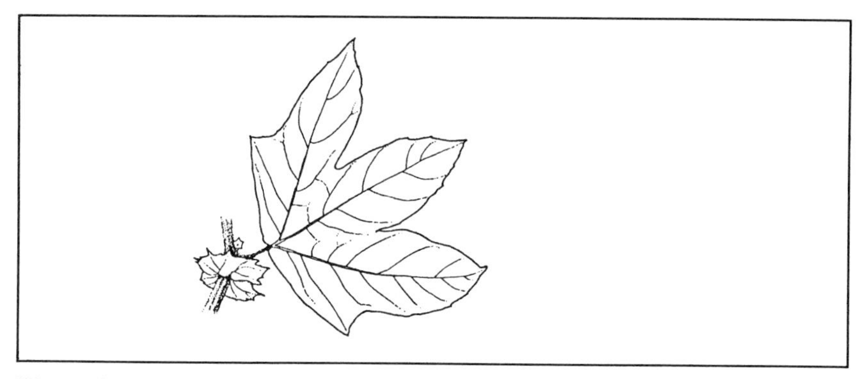

Western Sycamore *(Platanus racemosa)*

The California Walnut *(Juglans californica)* in the Walnut Family, the JUGLANDACEAE, is a small (15 to 30 feet tall) tree which often has several trunks branching from near the base. The distinctive leaves have 11 to 15 finely toothed, smooth leaflets. The smallish nut, edible and nutritious, is covered with a tough brown husk. The thick hulls, which produce a stain that can be avoided by wearing gloves, can be removed with a knife. This native tree has provided the sturdy rootstock for the commercial walnut crop in the state.

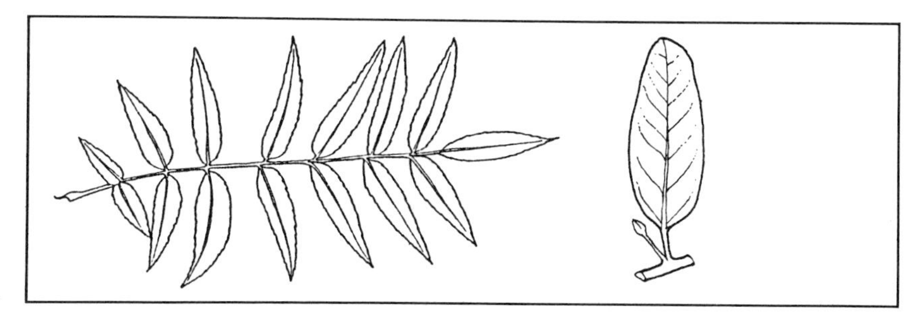

California Walnut *(Juglans californica)* California Bay *(Umbellularia californica)*

In the LAURACEAE is the handsome California Bay *(Umbellularia californica)* also known as Bay Laurel or Oregon Myrtle. It has a fine-grained, firm wood which takes a high polish and is carved into bowls and other souvenirs for sale in Oregon where it has been erroneously claimed that the species grows native only in that state and the Holy Land.

The California Bay, an evergreen tree, can reach a height of 90 feet. The lance-shaped leaves with smooth margins are thick, leathery and 3 to 5 inches long. When crushed, they have a pungent odor and are often used for flavor in cooking. The fruit is purple and fleshy, about an inch in diameter. Indians used the leaves to repel fleas and ate the nuts which they roasted to remove the bitter taste. The California Bay prefers a damp habitat along streams and moist slopes.

Big Leaf Maple *(Acer macrophyllum)* White Alder *(Alnus rhombifolia)*

Big Leaf Maple *(Acer macrophyllum)* of the Maple Family, the ACERACEAE, is a tall tree up to 100 feet with a trunk that can measure up to 30 inches in diameter. It is easily recognized by its large, 3 to 5 palmately lobed, shiny, dark green leaves which can be 18 inches long and as wide. The winged fruits, called samaras, are 1 to 2 inches long. Big Leaf Maple is found along streams and in steep side canyons.

In the Birch Family, the BETULACEAE, we have the White Alder *(Alnus rhombifolia)*, a graceful tree 30 to 100 feet tall with a straight trunk, grayish-brown patchy bark and spreading branches. The oval, toothed leaves are 2 to 4 inches long and up to 2 inches wide, dark green above and yellowish below. The leaf margins are not rolled under, which distinguishes it from the Red Alder *(A. oregona)* which is found further north. Our Alder produces small cones less than an inch long. White Alder is found along permanent streams below 1000 feet. Indians used it to make a red dye for their baskets and a tea to induce perspiration in their sweat houses.

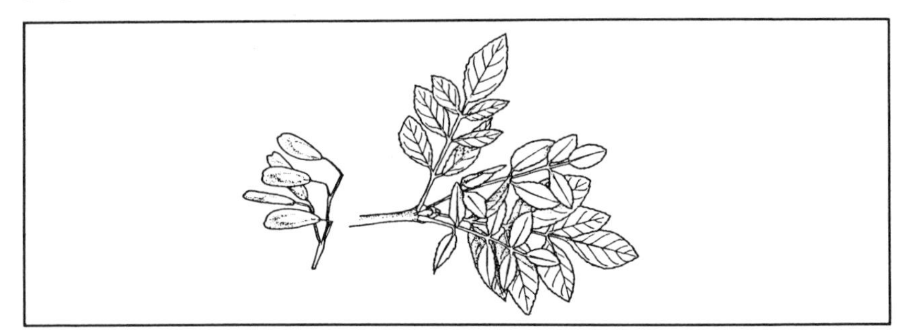

Flowering Ash *(Fraxinus dipetala)*

In our mountains, we have both the Flowering Ash *(Fraxinus dipetala)* and the Arizona Ash *(F. velutina)* which are members of the Olive Family, the OLEACEAE. They are both small, deciduous trees which prefer moist areas along streams and washes. Both have opposite pinnately compound leaves that are up to 6 inches long with 3 to 7 or 9 oval leaflets that are toothed above the middle. Although there are other differences, the best way to determine which is which is to examine the winged fruit, the samara. The samara of *F. dipetala* is winged along the sides, whereas the samara of *F. velutina* has a terminal wing.

In the Willow Family, the SALICACEAE, we have 2 cottonwoods and 3 willows. All of them occupy moist habitats along streams, wet places and the borders of salt marshes.

Fremont Cottonwood *(Populus fremontii)* and Black Cottonwood *(P. trichocarpa)* are both tall deciduous trees 40 to over 100 feet tall with broad open crowns. The leaves of *P. fremontii* are green on both surfaces while those of *P. trichocarpa* are distinctly whitish beneath. Both produce spike-like masses of flowers which are known as catkins. Male flowers are on one tree and female on another. The seeds are attached to fine, cottony hairs which float on the wind.

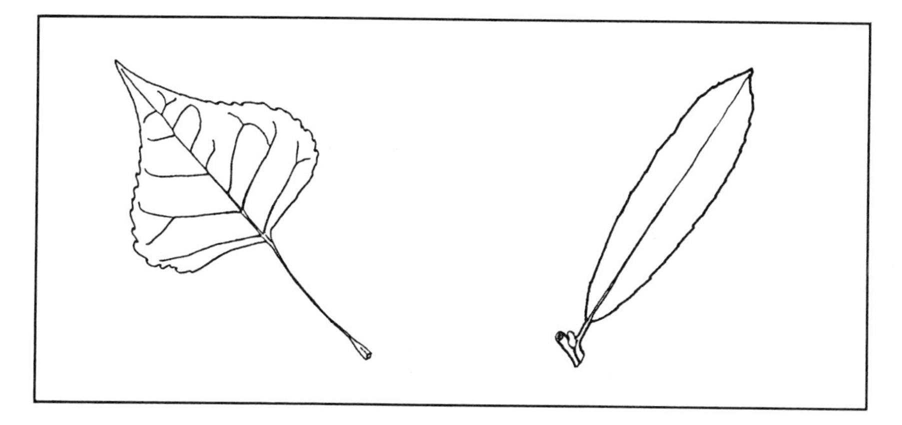

Fremont Cottonwood *(Populus fremontii)* Arroyo Willow *(Salix lasiolepis)*

In the genus *Salix,* the Santa Monicas have the Sandbar Willow *(S. hindsiana),* Arroyo Willow *(S. lasiolepis)* and Black Willow *(S. laevigata).* There are numerous species and varieties of willows in Southern California, many are more likely to be considered shrubs than trees. Since they hybridize freely, it is often very difficult to distinguish one from the other.

The leaves of willows are at least four times as long as they are wide which is not true of the cottonwoods. Sandbar Willow has such fine hairs on both sides of its leaves it appears silvery. Arroyo Willow has gray bark and catkins which look black. Black Willow has black bark and yellowish catkins.

Since all of the native trees of the Santa Monicas, except the oaks, inhabit moist locations, it is not surprising that there are not more of them. The Santa Monica Mountains are not a noticeably wet environment.

The mountains are surrounded by urban plantings so you may encounter species of eucalyptus, pepper trees, and Tree of Heaven *(Ailanthus altissima).* All are introduced. However, most are not more glorious than some of our natives.

NATIVE PLANT BOTANIC GARDENS

RANCHO SANTA ANA BOTANIC GARDEN, 1500 North College Ave., Claremont, CA 91711. 714/625-8767.
Eighty-three acres of natural plant communities with native California plants. Open Monday - Saturday 8-5; Sundays 10-5. Free.
Occasional sales of native plants.

SANTA BARBARA BOTANIC GARDEN, 1212 Mission Canyon Road, Santa Barbara, CA 93105. 805/682-4726.
Fifty acres of native trees, shrubs, wildflowers and cacti. Open daily 8 - Sunset. Free.
Occasional sales of native plants.

THEODORE PAYNE FOUNDATION, 10459 Tuxford Street, Sun Valley, CA 91352. 818/768-1802. Taped 24-hour wildflower report, 818/768-3533.
Devoted to the propagation of native plants for the home garden. Here you may learn how to grow native plants. Native plants may also be purchased. Open daily 9-4. Free.

SOUTH COAST BOTANIC GARDEN, 26300 Crenshaw Blvd., Palos Verdes Peninsula, CA 90274. 213/772-5813.
The garden contains over 2000 species, *not* all native plants. Open daily 9-5. Adults $1.50. Tram tours.

LUMMIS HOUSE, 200 East Avenue 43, Los Angeles, CA 90031. 213/222-0546.
Grounds with native plants. Open Wed.-Sun. 1-4. Free.

EARTHSIDE NATURE CENTER, Girls Club of Pasadena, 3160 East Del Mar Blvd., Pasadena, CA 91107. 818/796-6120.
Several acres of natives and propagated horticultural varieties built on an old neighborhood dump. Open some spring weekends. $2.

MILDRED E. MATHIAS BOTANICAL GARDENS, (UCLA) Hilgard and Le Conte Aves., Westwood, CA 90024. 213/825-4321.
Four thousand species of native and exotic plants are on eight acres. Self-guiding walking tours. Open Monday - Friday 8-5, Saturday - Sunday noon-5. Free.

LOS ANGELES STATE AND COUNTY ARBORETUM, 301 N. Baldwin Ave., Arcadia, CA 91006. 818/446-8251.
The 127 acres are filled with more than 30,000 plants, divided into geographical sections. Open daily 9-5. Adults $1.50. Tram tours.

DESCANSO GARDENS, south of Foothill Boulevard at 1418 Descanso Drive, La Canada, CA 91011. 818/790-5571.
The 166-acre garden includes a native plant section. Open daily 9-5. Adults $1.50. Tram tours.

HUNTINGTON BOTANICAL GARDENS, 1151 Oxford Road, San Marino, CA 91108. 818/792-6141.

A dozen botanical gardens, covering 130 acres of the 200-acre estate, feature 9000 plant, shrub and tree varieties. The desert gardens contains the largest outdoor collection in the world. Open Tuesday - Sunday, 1-4:30. Reservations needed on Sunday. Donation $2.00.

NATURE CLUBS

CALIFORNIA NATIVE PLANT SOCIETY, 909 12th Street, Suite 116, Sacramento, CA 95814, 916/447-CNPS (2677). A state wide organization with 26 chapters, several of which are located in Southern California, offering interpretive lectures, field trips, newsletters and a quarterly journal, *Fremontia.* One of these chapters is a local one—the SANTA MONICA MOUNTAINS CHAPTER, which, in addition to a bi-monthly newsletter, monthly programs, field trips and plant sales, sponsors a series of spring "Wildflower Walks" in the Santa Monica Mountains. For further information, write 6223 Lubao Ave., Woodland Hills, CA 91367.

SOUTHERN CALIFORNIA BOTANISTS, Rancho Santa Ana Botanic Garden, 1500 North College Avenue, Claremont, CA 91711. Founded in 1927. Members include professional botanists and interested laypersons. Activities include field trips, an annual symposium, lectures and a potluck. Their journal, *Crossosoma,* contains articles of both scientific and general interest.

SIERRA CLUB, Angeles Chapter, Natural Science Section, 2410 W. Beverly Blvd., Los Angeles, CA 90057. 213/387-4287. Holds evening programs of illustrated lectures, field trips and publishes a monthly newsletter.

SIERRA CLUB, Santa Monica Mountains Task Force, 2410 W. Beverly Blvd., Los Angeles, CA 90057. 213/387-4287. Organizes and conducts "Sundays in the Santa Monicas", a series of hikes (not always on Sundays) ranging from easy strolls to the very vigorous. Write to the above address for hike schedule. Free to the public. They also publish a most useful booklet called *Day Walks in the Santa Monicas.* To order, send $4.50 to: Sierra Club, 4961 Edgerton Ave., Encino, CA 91316.

DOCENT ORGANIZATIONS

TOPANGA CANYON	818/789-3456
WILLIAM O. DOUGLAS OUTDOOR CLASSROOM (WODOC)	213/858-3843
MALIBU CREEK STATE PARK	818/706-1310
COLD CREEK CANYON	213/456-5627
CHARMLEE PARK	213/457-7247
WILL ROGERS STATE HISTORICAL PARK	213/454-8212

PUBLIC AGENCIES

THE SANTA MONICA MOUNTAINS NATIONAL RECREATION AREA (SMMNRA), 22900 Ventura Blvd., #140, Woodland Hills, CA 91364. 818/888-3770. The National Park Service is one of several agencies that administer and provide recreation programs within the boundaries of the SMMNRA.

CALIFORNIA DEPARTMENT OF RECREATION AND PARKS, Santa Monica Mountains Area Office, 2860-A Camino Dos Rios, Newbury Park, CA 91320. 818/706-1310 and 805/987-3303.

LOS ANGELES COUNTY DEPARTMENT OF PARKS AND RECREATION, 19152 W. Placerita Canyon Road, Newhall, CA 91321. 805/259-7721.

LOS ANGELES CITY RECREATION AND PARKS DEPARTMENT, Room 1330, City Hall East, Los Angeles, CA 90012. 213/485-5515.

SANTA MONICA MOUNTAINS CONSERVANCY, 107 South Broadway, Room 7117, Los Angeles, CA 90012. 213/620-2021.

MOUNTAINS RESTORATION TRUST, 21361 West Pacific Coast Highway, Malibu, CA 90265. 213/456-5625.

CONEJO PARKS AND RECREATION DISTRICT, 401 W. Hillcrest Drive, Thousand Oaks, CA 91360. 805/495-6471.

BIBLIOGRAPHY

Here are some books that will help you in your further studies of wildflowers:

Abrams, Leroy, and Ferris, Roxana S. *ILLUSTRATED FLORA OF THE PACIFIC STATES.* Stanford, Stanford University Press, 1923-1960.

This is a four-volume set with 2,771 pages covering thousands of species. Each species is illustrated in black and white. There are keys.

Collins, Barbara J. *KEY TO COASTAL AND CHAPARRAL FLOWERING PLANTS OF SOUTHERN CALIFORNIA.* Northridge, California State University Foundation, 1970.

This 249-page, inexpensive paperback is an excellent reference for the beginning enthusiast. The keys, the organization of the sections, the line drawings that picture each plant and each botanical term are all designed to be helpful—which they are indeed to the amateur.

Dr. Collins has since published three additional paperback books, all similarly organized. Two of these are specific for the deserts of Southern California. The other is listed below.

Collins, Barbara J. *KEY TO TREES AND WILDFLOWERS OF THE MOUNTAINS OF SOUTHERN CALIFORNIA.* Northridge, California State University Foundation, 1974.

This contains 277 pages.

Fitter, Richard, and Fitter, Alastair. *THE WILD FLOWERS OF BRITAIN AND NORTHERN EUROPE.* Collins, St. Jame's Place, London, 3rd edition, 1978.

This beautifully illustrated (color and black and white) book has 336 pages and expands our knowledge of alien plants.

Harrington, H.D., *HOW TO IDENTIFY PLANTS.* Ohio University Press, Athens, Ohio, 1957.

This 203-page paperback deals with the technical terms used in botany and their definitions complete with line drawings. Each chapter explains terms as related to a specific part of plants, that is, terms relative to the leaves, fruits and seeds, etc. A truly valuable aid to learning the language of botany.

Mason, Herbert L. *A FLORA OF THE MARSHES OF CALIFORNIA.* University of California Press, Berkeley and Los Angeles, 1957.

This 878-page volume is strongly recommended for those who become interested in the often difficult to identify plants of the wetlands. It is very thorough and the numerous line drawings are explicit and helpful.

McMinn, Howard E. *AN ILLUSTRATED MANUAL OF CALIFORNIA SHRUBS.* University of California Press, Berkeley and Los Angeles, 1979.

This is a well written and well illustrated 663-page guide to the many native shrubs of the state.

Munz, Philip A., and Keck, David D. *A CALIFORNIA FLORA.* University of California Press, Berkeley, 1959. With SUPPLEMENT, 1968.

This is the definitive work to date on the flora of the state. It is quoted, cited and consulted for final decisions. It is a difficult book (1,681 pages) for a beginner but essential for anyone with a serious interest.

Munz, Philip A. *A FLORA OF SOUTHERN CALIFORNIA.* University of California Press, Berkeley, 1974.

This was the last book Dr. Munz wrote. He died just after its publication. It is easier to use than the one that attempts to include all the native plants of the state because by definition it contains far fewer species. It also contains many more black and white illustrations which are always helpful. There are 1,086 pages.

Niehaus, Theodore F. *A FIELD GUIDE TO PACIFIC STATES WILDFLOWERS.* Houghton Mifflin Company, Boston, 1976.

This Peterson Field Guide Series which uses the visual approach to color, form and detail of wildflowers is an excellent book to carry along on field trips. It contains 1,492 species grouped by color with family keys and points out the features that distinguish one flower from another. There are 432 pages.

Parson, Mary E. *THE WILDFLOWERS OF CALIFORNIA.* Dover Publications, New York, 1966.

This is an updated edition of a delightful work first published in 1897. There are numerous, full-page, pen and ink drawings. Families and important genera are described as well as hundreds of species which are arranged by color. The key is based on stamens. There are 425 pages.

Raven, Peter H., and Thompson, Henry J. *FLORA OF THE SANTA MONICA MOUNTAINS, CALIFORNIA.* University of California, Los Angeles, 1966. Revised 1974. Updated revision to be published soon.

This inexpensive paperback is an unusually well done local flora. It is a comprehensive and authoritative guide to all the vascular plants of the range. It is not illustrated but the simple keys are most explicit. There are 189 pages.

Stearn, William T. *A GARDENER'S DICTIONARY OF PLANT NAMES.* St. Martins Press, New York, 1972.

University of California Press publishes a series of "California Natural History Guides" which are all reasonably priced, in paperback and available in most bookstores. The following were most helpful in compiling this book and will certainly be useful to anyone who has developed an interest in the flora of the Santa Monica Mountains.

California Insects, Powell and Hogue
Early Uses of California Plants, Balls
Edible and Useful Plants of California, Clark
Growing California Native Plants, Schmidt
Introduction to California Plant Life, Ornduff
Native Shrubs of Southern California, Raven
Native Trees of Southern California, Peterson
Seashore Plants of California, Dawson and Foster

UPDATING SCIENTIFIC NAMES

Botanical names undergo constant change. The nomenclature in this book agrees with the nomenclature in *A Flora of Southern California* by Philip A. Munz. However, with time and research, certain changes are made in classification. The following changes were made in the 1986 revision of the *Flora of the Santa Monica Mountains, California.* It is expected that by 1900 several other significant changes will have been made.

OLD NAME	*NEW NAME*
Rhus laurina	*Malosma laurina*
Rorippa nasturtium-aquaticum	*Nasturtium officinale*
Isomeris arborea	*Cleome isomeris*
Salsola iberica	*Salsola australis*
Euphorbia albomarginata	*Chamaesyce albomarginata*
Clarkia deflexa	*Clarkia bottae*
Zauschneria californica	*Epilobium canum* ssp. *canum angustifolium*
Zauschneria cana	*Epilobium canum* ssp. *canum*
Datura meteloides	*Datura innoxia*

GLOSSARY

These flowers have been described in the simplest words possible to insure that with a little interest and application and a hard look at the pictures, you can identify them. It hasn't been possible to totally eliminate some of the terms that might confuse you. So these are further explained here.

Achene, a small dry 1-seeded fruit, which does not open at maturity.
Alternate, one side and then the other; not opposite or whorled; said of a single leaf or branch at a node.
Annual, grows from seed, flowers, produces seeds and dies in one year.
Anther, upper portion of the stamen containing the pollen.
Apex, tip.
Appressed, flattened against or close, said of hairs or pods.
Axil, angle between the leaf and the stem.
Basal, occurring at the base.
Beak, the long, narrow tip on a fruit or the projection on the keel of some legume flowers.
Berry, a fleshy, pulpy fruit with immersed seeds lacking a true stone; not splitting at maturity.
Biennial, grows from seed to maturity and dies in two years.
Blade, wide part of petal or leaf.
Bract, a leaf-like structure associated with a flower cluster.
Bulb, underground storage structure composed of stem tip and fleshy leaves.

Bur, a fruit with spines, usually hooked or barbed.
Calyx, collective term for the sepals which may be separate or united.
Capitate, in a dense head-shaped cluster.
Capsule, a dry fruit from a compound ovary, opening at maturity.
Catkin, a pendulous spiked cluster of unisexual flowers.
Clasping, base of leaf surrounds or partly surrounds the stem.
Claw, the narrowed base of some petals; the expanded portion is the blade.
Compound, composed of more than one similar part.
Compound leaf, a leaf completely separated into two or more leaflets.
Corolla, collective term for the petals of a flower which may be separated or united.
Dicot, a plant with two seed leaves and net-veined leaves.
Disk floret, a type of flower in the ASTERACEAE.
Entire, smooth-margined, without teeth or lobes.
Exserted, protruding beyond, as stamens protruding beyond the corolla.
Family, grouping of plants made up of one or more related genera.
Follicle, a dry fruit that splits open lengthwise along one line.
Fruit, matured ovary of a flower; contains the seeds.
Genera, plural of genus.
Genus, grouping of plants made up of a number of related species.
Habitat, place where the plant lives.
Head, dense cluster of flowers or fruits without stalks, used especially for the ASTERACEAE.
Herb, a plant without a woody stem.
Herbage, plant stems and leaves.
Herbaceous, like an herb.
Imbricate, partly overlapping like shingles on a roof.
Inferior, see ovary.
Inflorescence, cluster of flowers.
Lance-like, several times longer than wide, broadest toward the base, tapering to the tip.
Leaflet, one of the divisions of a compound leaf.
Legume, fruit of the Pea Family, also said of any plant of this family.
Lobe, division of a leaf or a petal usually rounded.
Mealy, a surface covered with whitish, granular particles that can be rubbed off.
Monocot, a plant with one seed leaf and parallel veined leaves.
Nerve, a simple, unbranched vein or a slender rib in a leaf.
Node, the place on a stem where a leaf or a branch arises or is attached.
Nut, a 1-seeded fruit with a hard wall.
Nutlet, a small nut-like fruit of LAMIACEAE, BORAGINACEAE, and ZYGOPHYLLACEAE.
Opposite, one on each side of a node; paired on opposite sides of the stem.
Ovary, swollen basal portion of the pistil; said to be inferior when it is below the petals; said to be superior when it is above the petals.
Ovate, egg-shaped in outline.
Palmate, finger-like; fanning out from a common point as in leaves.
Pappus, the various tufts of hairs, bristles or scales on achenes or fruits.
Parasite, an organism that grows on and obtains nourishment from another; usually lacking chlorophyll in plants.
Parted, lobed or cut-in over halfway, as in a leaf.

Perennial, living for many years.
Perianth, the envelope of the flower which surrounds the stamens and/or pistil(s); it may consist of sepals and petals or of sepals only.
Petals, the separate or united parts of the corolla.
Pinnate, feather-like; arranged on either side of a main stem as in leaflets.
Pistil, female portion of the flower; will mature into fruit.
Pod, any dry fruit that splits open on two sides when mature; legume-like.
Pollen, male reproductive bodies found in the anther of the stamen; a fine, powdery substance.
Prostrate, lying flat on the ground.
Ray floret, a type of flower in the ASTERACEAE.
Receptacle, the base of the flower.
Rosette, a dense, basal cluster of leaves.
Samara, a dry, winged fruit.
Schizocarp, a dry fruit splitting at maturity into two or more one-seeded segments which do not open.
Segment, part or portion, as of a perianth or divided leaf.
Sepal, usually green, leafy portion of a flower; occurs beneath the petals.
Shrub a woody plant smaller than a tree.
Simple, unbranched; of one piece; not compound.
Species, major subdivison of a genus; related individuals resemble each other; able to reproduce among themselves.
Spike, an elongated inflorescence of flowers without individual stalks.
Stamen, the male portion of the flower; produces the pollen.
Sterile, without male or female organs.
Stigma, the top portion of the pistil which receives the pollen.
Stipules, small leaf-like appendages found at the base of a leaf or its stalk.
Style, upper, narrow neck of the pistil below the stigma and above the ovary.
Succulent, thick, fleshy, juicy.
Superior, see ovary.
Tendril, spiral structure arising from the stem or leaf; enables climbing.
Terminal, at the end of the stem.
Tooth, small, rounded or pointed lobe on the edge of a leaf or a petal.
Tubular, shaped like a hollow cylinder.
Two-lipped, said of a bilaterally symmetrical flower with upper and lower lobes.
Umbel, a flower cluster composed of flowers on stalks that rise at the summit of a stem.
Veins, ridges in a leaf forming framework; vascular bundles.
Whorl, an arrangement of 3 or more parts radiating from a node.
Winged, provided with thin papery wings or extensions.
Woolly, with long, soft, interwoven hairs.

INDEX TO FLOWERS BY COLOR

Blue, Purple, Lavender, Red-Violet

Brown

Pink

Red

White, Cream, Greenish-white

Yellow or Orange

LISTING OF PLANTS BY FAMILY

MONOCOTS

AGAVACEAE, Agave Family, 23-24
Yucca whipplei, Our Lord's Candle, 23-24

AMARYLLIDACEAE, Amaryllis Family, 24
Allium peninsulare, Peninsular Onion, 24
*Allium haematochiton,** Red-skinned Onion,* 25
Bloomeria crocea, Golden Stars, 25
Brodiaea jolonensis, Harvest Brodiaea, 26
Dichelostemma puchellum, Blue Dicks, Wild Hyacinth, 26

IRIDACEAE, Iris Family, 27
Sisyrinchium bellum, Blue-eyed grass, 27

LILIACEAE, Lily Family, 28
Calochortus albus, Fairy Lantern, Globe Lily, 28
Calochortus catalinae, Catalina Mariposa Lily, 29
Calochortus clavatus, Yellow Mariposa Lily, 30
Calochortus plummerae, Pink Mariposa Lily, 31
Calochortus splendens, Splendid Mariposa Lily, 31
Calochortus venustus, Butterfly Mariposa, 32
Chlorogalum pomeridianum, Soap Plant, Amole, 33
Fritillaria biflora, Chocolate Lily, 34
Lilium humboldtii, Humboldt Lily, 34
Zigadenus fremontii, Star Lily, Fremont Zigadene, 35
*Zigadenus venenosus**, Death Camas,* 35

ORCHIDACEAE, Orchid Family, 36
Epipactis gigantea, Stream Orchid, 36
Habenaria unalascensis, Rein Orchid, 37

DICOTS

AIZOACEAE, Carpetweed Family, 38
Carpobrotus aequilaterus, Sea Fig, 38
Carpobrotus edulis, Hottentot Fig, 39
Mesembryanthemum crystallinum, Common Ice Plant, 40
*Mesembryanthemum nodiflorum**, Slender-leaved Ice Plant,* 40
Tetragonia tetragonioides, New Zealand Spinach, 40

ANACARDIACEAE, Sumac Family, 41
Rhus integrifolia, Lemonadeberry, 41
Rhus laurina, Laurel Sumac, 42
Rhus ovata, Sugar Bush, 43
Rhus trilobata, Squaw Bush, 43
Toxicodendron diversilobum, Poison Oak, 44

APIACEAE (UMBELLIFERAE), Carrot Family, 45
Apium graveolens, Common Celery, 45
Conium maculatum, Poison Hemlock, 46
Foeniculum vulgare, Sweet Fennel, 47
Lomatium spp., 48
*Pimpinella anisum,** Anise Seed,* 47
Sanicula arguta, Sharp-toothed Sanicle, 48
Tauschia spp.,* 48

ASCLEPIADACEAE, Milkweed Family, 49
Asclepias fascicularis, Narrow-leaved Milkweed, 49

ASTERACEAE (COMPOSITAE), Sunflower Family, 49
Achillea millefolium, White Yarrow, Milfoil, 50
*Ambrosia acanthicarpa,** 51
Ambrosia chamissonis , Chamisso's Burweed, 51
*Ambrosia confertiflora,** 51
Ambrosia psilostachya, Western Ragweed, 51
*Artemisia absinthium,** Absinthe,* 52
Artemisia californica, Coastal Sagebrush, 52
Artemisia douglasiana, California mugwort, 52
*Artemisia dracunculus,** Tarragon,* 52
Baccharis glutinosa, Mule Fat, 53
Baccharis pilularis, Coyote Bush, 54
Bidens pilosa, Hairy Bur-marigold, 55
Brickellia californica, California Brickelbush, 55-56
Centaurea melitensis, Yellow Star Thistle, Tocalote, 56
Chaenactis artemisiaefolia, White Chaenactis, 57
*Chaenactis fremontii,** Fremont Pincushion,* 57
Chaenactis glabriuscula, Yellow Chaenactis, 57
*Cichorium intybus,** 72
*Cirsium californicum,** 58
Cirsium occidentale, Western Thistle, Cobweb Thistle, 58
*Cirsium vulgare,** Bull Thistle,* 58
Coreopsis bigelovii, Bigelow Coreopsis, 58
*Coreopsis douglasii,** 59
Coreopsis gigantea, Giant Coreopsis, Sea Dahlia, 59
Corethrogyne filaginifolia, Common Corethrogyne, Cudweed-aster, 60
Cotula coronopifolia, Brass Buttons, 60
Encelia californica, California Encelia, Bush Sunflower, 61
*Encelia farinosa,** Incienso,* Brittle-bush,* 61
Erigeron foliosus, Leafy Daisy, Fleabane, 62

TREES

**Not illustrated.*

GENERAL INDEX

**Not illustrated.*

ABOUT THE AUTHOR

Nancy Dale, a Fellow of the California Native Plant Society, credits the Society for her knowledge of and continuing interest in the native flora of the state. She has been a member since its founding 20 years ago in 1965 and much of her botanical knowledge has come from the professionally led field trips and slide/lecture programs the society continues to offer.

She was president of the Santa Monica Mountains Chapter for several years and edited the chapter newsletter before, during and after her presidency. She continues to write a column for this publication entitled "Featured Family" which explains in non-technical terms groups of plants found in the mountains. She leads flower walks in the Santa Monicas interpreting the native vegetation to the general public, and she is available as a speaker.

Articles she has written have appeared in *Fremontia,* the quarterly journal of The California Native Plant Society, and *Pacific Horticulture.*